The
Blood–Brain Barrier
and Drug Delivery
to the CNS

The
Blood–Brain Barrier
and Drug Delivery
to the CNS

edited by

David J. Begley
King's College London
London, England

Michael W. Bradbury
King's College London
London, England

Jörg Kreuter
Biozentrum-Niederursel
Johann Wolfgang Goethe-Universität
Frankfurt am Main, Germany

CRC Press
Taylor & Francis Group
Boca Raton   London   New York

CRC Press is an imprint of the
Taylor & Francis Group, an **informa** business

CRC Press
Taylor & Francis Group
6000 Broken Sound Parkway NW, Suite 300
Boca Raton, FL 33487-2742

First issued in paperback 2019

© 2010 by Taylor & Francis Group, LLC
CRC Press is an imprint of Taylor & Francis Group, an Informa business

No claim to original U.S. Government works

ISBN-13: 978-0-8247-0394-3 (hbk)
ISBN-13: 978-0-367-39866-8 (pbk)

A CIP record for this book is available from the British Library.

Library of Congress Cataloging-in-Publication Data available on application

**Visit the Taylor & Francis Web site at
http://www.taylorandfrancis.com**

**and the CRC Press Web site at
http://www.crcpress.com**

Preface

Now that the search for and development of new drugs are increasingly based on models relating function to structure, it is becoming apparent that in the case of drugs acting on the central nervous system (CNS), such models are also crucial for the mechanisms determining uptake into the brain. This realization implies better understanding of the relationship of transport at the blood–brain barrier to drug structure and physicochemical properties. Within this volume we have included a number of chapters describing crucial current areas of investigation relevant to CNS drug penetration. These include mathematical and computer models to predict drug entry and distribution in the brain on the basis of molecular descriptors and the use of various in vitro models of the blood–brain barrier, composed of cultured endothelial cells, to further predict transport of drugs and solutes.

The potential of microdialysis in studying the kinetics of drug entry into the CNS and in determining the true brain interstitial fluid concentration of drugs is explored. Three chapters deal with specific transporters for drugs into and out of the brain, with special emphasis on the rapidly expanding area of multidrug transporters and their role in restricting CNS penetration of drugs. Other chapters examine the metabolism of drugs within the brain and the mechanisms of activation and detoxification of pharmacological agents; the targeting of large molecules, including proteins and peptides; and the new technique of introducing drugs conjugated with nanoparticle suspensions. Finally, the possibilities of using imaging techniques to record drug kinetics and binding in the living human subject are introduced.

This volume will prove invaluable to researchers interested in the fundamental function of the blood–brain barrier and to those in the pharmaceutical industry interested in rational drug design directed at targeting drugs to the brain.

This book arose from a symposium with the same title, held in Frankfurt am Main, Germany. The symposium was organized by the editors during a period when Michael Bradbury and David Begley jointly held the Friedrich Merz Visiting Professorship at the Institute of Pharmaceutical Technology, University of Frankfurt. This Chair is endowed by the Merz Pharmaceutical Company and the authors are grateful to the company and Dr. Jochen Hückman, who gave much support to the symposium and to us during our stay in Germany. Thanks are also due to Johann Wolfgang Goethe-Universität and to Professor Kreuter's Institute for hosting the symposium and making our time in Frankfurt a happy and productive occasion.

In the interests of fully comprehensive coverage of the current field we have invited a few additional scientists of international stature to contribute. We thank the contributors for their time and effort in producing this book.

David J. Begley
Michael W. Bradbury
Jörg Kreuter

Contents

Contributors

Joan Abbott   Division of Physiology, King's College London, London, England

Michael H. Abraham   Department of Chemistry, University College London, London, England

Renad N. Alyautdin   Department of Pharmacology, I. M. Sechenov Moscow Medical Academy, Moscow, Russia

Claire Bayol-Denizot   Département de Pharmacologie, Université Henri Poincaré-Nancy 1, CNRS, Vandoeuvre-lès-Nancy, France

David J. Begley   Division of Physiology, King's College London, London, England

Ulrich Bickel   Department of Pharmaceutical Sciences, School of Pharmacy, Texas Tech University, Amarillo, Texas

Michael W. Bradbury   Division of Physiology, King's College London, London, England

Douwe D. Breimer   Division of Pharmacology, Leiden/Amsterdam Center for Drug Research, University of Leiden, Leiden, The Netherlands

V. Buée-Scherrer   Laboratoire de Biologie Cellulaire et Moléculaire, Université d'Artois, Lens, France

R. Cecchelli   Laboratoire de Biologie Cellulaire et Moléculaire, Université d'Artois, Lens, France

A. G. de Boer   Division of Pharmacology, Leiden/Amsterdam Center for Drug Research, University of Leiden, Leiden, The Netherlands

Elizabeth C. M. de Lange   Division of Pharmacology, Leiden/Amsterdam Center for Drug Research, University of Leiden, Leiden, The Netherlands

B. Dehouck   INSERM U325, Institut Pasteur, Lille, France

M. P. Dehouck   INSERM U325, Institut Pasteur, Lille, France

L. Descamps   INSERM U325, Institut Pasteur, Lille, France

C. Duhem   INSERM U325, Institut Pasteur, Lille, France

Ramon D. S. El-Bachá   Département de Pharmacologie, Université Henri Poincaré-Nancy 1, CNRS, Vandoeuvre-lès-Nancy, France

Christiane Engelbertz   Institute of Biochemistry, University of Muenster, Muenster, Germany

L. Fenart   Laboratoire de Biologie Cellulaire et Moléculaire, Université d'Artois, Lens, France

Helmut Franke*   Institute of Biochemistry, University of Muenster, Muenster, Germany

Hans-Joachim Galla   Institute of Biochemistry, University of Muenster, Muenster, Germany

Antony D. Gee   Smith Kline Beecham, CRU, ACCI, Addenbrookes Hospital, Cambridge, England

---

* *Current affiliation*: Hoffmann LaRoche, Basel, Switzerland

Albert Gjedde Positron Emission Tomography Center, Aarhus General Hospital, Aarhus, Denmark

Daniela Gradinaru National Institute of Gerontology and Geriatrics, Ana Aslan, Bucharest, Romania

Matthias Haselbach Institute of Biochemistry, University of Muenster, Muenster, Germany

Ehsan Ullah Khan Florence Nightingale School of Nursing and Midwifery, King's College London, London, England

Dorothea Korte Institute of Biochemistry, University of Muenster, Muenster, Germany

Jörg Kreuter Institut für Pharmazeutische Technologie, Biozentrum-Niederursel, Johann Wolfgang Goethe-Universität, Frankfurt am Main, Germany

Philippe Lagrange Département de Pharmacologie, Université Henri Poincaré-Nancy 1, CNRS, Vandoeuvre-lès-Nancy, France

Ulrich Mayer Department of Oncology and Hematology, University Clinic Eppendorf, Hamburg, Germany

Alain Minn Département de Pharmacologie, Université Henri Poincaré-Nancy 1, CNRS, Vandoeuvre-lès-Nancy, France

Thorsten Nitz Institute of Biochemistry, University of Muenster, Muenster, Germany

James A. Platts Department of Chemistry, University College London, London, England

Anthony Regina INSERM U26, Unité de Neuro-Pharmaco-Nutrition, Hôpital Fernand Widal, Paris, France

Christopher Rollinson Division of Physiology, King's College London, London, England

Françoise Roux   INSERM U26, Unité de Neuro-Pharmaco-Nutrition, Hôpital Fernand Widal, Paris, France

Donald Frederick Smith   Department of Biological Psychiatry, Institute for Basic Research in Psychiatry, Psychiatric Hospital, Risskov, Denmark

Funmilayo G. Suleman   Département de Pharmacologie, Université Henri Poincaré-Nancy 1, CNRS, Vandoeuvre-lès-Nancy, France

G. Torpier   INSERM U325, Institut Pasteur, Lille, France

Akira Tsuji   Department of Pharmacobio-Dynamics, Kanazawa University, Kanazawa, Ishikawa, Japan

Inez C. J. van der Sandt   Division of Pharmacology, Leiden/Amsterdam Center for Drug Research, University of Leiden, Leiden, The Netherlands

Joachim Wegener*   Institute of Biochemistry, University of Muenster, Muenster, Germany

---

* *Current affiliation*: Rensselaer Polytechnique Institute, Troy, New York

1
History and Physiology of the Blood–Brain Barrier in Relation to Delivery of Drugs to the Brain

Michael W. Bradbury
King's College London, London, England

I. INTRODUCTION

Among the factors that influence uptake and distribution of a drug into a tissue or organ outside the central nervous system (CNS) are blood flow, binding to proteins in blood plasma, clearance from blood, and metabolism. Since capillaries in general have numerous water-filled channels across their walls which allow ready escape of solutes up to the size of small proteins into interstitial fluid, capillary permeability is not normally a factor in limiting drug uptake. The contrary is true for brain and spinal cord. The blood–brain barrier (BBB) is frequently a rate-limiting factor in determining permeation of a drug into brain and is also an element in influencing steady state distribution. It is commonplace for pharmacologists and pharmaceutical chemists to make predictions concerning the pharmacological activity of a potential drug from its molecular structure. Those administering drugs active on cells of the CNS are generally concerned to raise and maintain an optimum concentration of the agent in the cerebral interstitial fluid where it is in contact with receptors. It would be most helpful to be able to predict both permeability at the BBB and steady state distribution of a potential neuropharmacological agent from its molecular structure.

II. IDEAS RELATING BRAIN UPTAKE TO PHYSICOCHEMICAL PROPERTIES OF DRUGS

An early review that considers the physical chemistry of molecules in relation to brain entry is that of Friedemann (1). While this author accepted a role for lipid solubility, his thesis is almost entirely devoted to the view that molecular charge is crucial for penetration of non-lipid-soluble compounds, cationic molecules being permeant and anionic nonpenetrating. However, it was Krogh (2), in his great survey of exchanges through the surface of living cells and across living membranes generally, who concluded that in the search for drugs that act on the CNS one should be guided by a substance's solubility in lipids rather than by its electrical charge. He indicated that the cerebral capillaries have the general properties of the cell membrane and that they may even have secretory properties (i.e., be capable of active transport). Over half a century ago, Krogh foresaw much of what we now know to be true of the cerebral endothelium.

Davson (3) related rate of exchange in cerebrospinal fluid (CSF) and brain to molecules in a series, including thiourea itself and a range of alkyl-substituted thioureas. The rate clearly varied with the degree of hydrocarbon insertion (i.e., also with the solubility in lipid solvents). Later, it became fashionable to compare rate of exchange in CSF or in CSF and brain with the measured partition coefficient of the compound between a fat solvent and water. Thus Rall et al. (4) matched rate of exchange of each of a series of sulfonamides with its respective nonionized fraction and to its chloroform–water partition coefficient. Mayer and colleagues (5) compared rates for a different series of drugs in CSF and brain with the respective partition coefficients in heptane, benzene, and chloroform against water. More recently it has become usual to relate the logarithm of brain permeation into brain with the logarithm of octanol–water partition coefficient log P_{oct} (e.g., Rapoport and Levitan, 6). This change in use of lipid solvent as a standard of comparison seems to have been based on studies of the pharmacological activities of molecules on the brain rather than on measured penetrations (7). These very different measurements may or may not give comparable results. No rigorous studies of the value of different lipid solvents in predicting BBB permeation from partitioning seem to have been made. All lipid solvents mentioned seem to yield tolerable comparisons, but inspection of regressions plotted in the literature of large groups of compounds suggests that measured values of permeability may in some instances differ from predicted values by as much as an order of magnitude.

III. RELATION OF BRAIN UPTAKE TO MOLECULAR STRUCTURE

A more fundamental approach is to directly relate measured permeability of the blood–brain barrier or steady state distribution of a compound between blood and brain to molecular factors. Thus the linear free energy relationship permits distribution of dissolved molecules between phases to be related to the quantitative influence of certain solute molecular "descriptors" (8). These descriptors include excess molar refraction, dipolarity/polarizability, overall hydrogen bonding acidity and basicity, respectively, and the characteristic volume of McGowan. The method has been applied to both blood–brain distribution (9) and to permeability at the BBB (10). In both cases the method provided substantially better predictions of solute behavior than did the octanol–water partition coefficient. The approach is reviewed in much greater detail elsewhere in this volume (11).

The partial success of partition coefficients and the greater precision of the Abraham relation in predicting permeability of the BBB must depend on the dependence of rate of passive diffusion across the cerebral endothelium on the level of solute partitioning in the lipid plasma membranes of the endothelial cells. Hence, it is evident that where other specific processes are involved in enhancing or in restraining movement of molecules across the endothelium, this prediction on its own will underestimate or overestimate, respectively, the true permeability. The number of observed exceptions to molecular transport by passive diffusion alone are increasing and are detailed in several later chapters.

IV. SPECIFIC TRANSPORT MECHANISMS AT THE BLOOD–BRAIN BARRIER

A number of nutrient molecules are transferred across the BBB by "facilitated diffusion." The transported molecules cross the plasma membranes by interacting with intramembrane transporter proteins related to water-filled channels. The two systems with the highest capacity are that for D-glucose and certain other sugars (the gene product *Glut 1*) and that for large neutral amino acids, the so-called L-system. The first has been sequenced for a number of mammalian species, the preferred structure of the sugar substrate understood, and models for sugar translocation across the membrane discussed (12). The high maximum transport capacity at the blood–brain barrier, 4 μmol

$min^{-1}g^{-1}$ in the rat and 1 µmol $min^{-1}g^{-1}$ in man (13), suggests that this system might be used for transport into brain of a drug linked to a D-pyranose sugar of appropriate structure.

The L transporter at the blood–brain barrier has somewhat different K_m values for its substrate amino acids than the K_m values exhibited in other tissues; hence it is regarded as a separate isoform and has been designated L1 (14). The general L transporter has recently been sequenced (15) and the three-dimensional structure of the binding site for neutral amino acids at the blood–brain barrier has been largely established by computer modeling (16). Marked preference for phenylalanine analogues was exhibited when a neutral substituent was at the meta position. The anticancer drugs melphalan and D,L-2-NAM-7 are appreciably transported by the L1 process. The latter drug has an exceptionally high affinity for the transporter, with a K_m of about 0.2 µM (17). The transporter also has pharmaceutical significance in that it carries L-Dopa, used in the therapy of Parkinson's disease.

The transporter for basic amino acids is also of interest in that the gene has been cloned for the murine (18), rat, and human system. The V_{max} is rather less than that for the L1 transporter, at 24 nmol $min^{-1}g^{-1}$ (14). With the advent of therapies directed toward DNA, a further system of transporters of relevance is that with affinity for certain nucleosides analogues. The components are being characterized in various tissues (19), but that at the blood–brain barrier (20) deserves further research.

It is becoming increasingly apparent that active transport plays an important role in restricting the entry of certain drugs into brain. The function of active transport in pumping certain organic anions, such as p-aminohippurate, from CSF to blood has been recognized since the early 1960s (21). The choroid epithelium is particularly effective in this activity, but it also occurs at the cerebral endothelium (22). The system is now called the multispecific organic anion transporter (23) and is effective in removing penicillin and azidothymidine (AZT) from CSF and brain (24). The efflux is blocked by probenecid.

It had been supposed that there might be a molecular weight limit on drug entry into brain from blood, uptake of drugs above about 800 kDa being small (25). However, drugs Levin studied, such as doxorubicin, vincristine, and etopside, are now known to be substrates for a potent mechanism that restricts brain entry of a wide range of drugs. Thus P-glycoprotein is sited in the apical (blood-facing) plasma membrane of the endothelium and is able to utilize ATP in pumping certain drugs from endothelial cells into blood, thus reducing brain entry, as reviewed by Borst and Schinkel (26) and in this volume by Begley et al. (23) and by Mayer (27). Substrates include vinblastine, ivermectin, digoxin, and cyclosporine. Molecules transported are often lipophilic and larger in size than many drugs. The structures of transported molecules differ markedly, and the structural characteristics of the binding site have not yet been fully defined.

Movement into or through the cerebral endothelium of molecules of the size of peptides and proteins (e.g., insulin, transferrin, lipoproteins) may depend on receptor-mediated endocytosis. Alternatively, cationization of albumin may stimulate a nonspecific endocytosis. Significant flux into brain of drugs linked to monoclonal antibodies against the transferrin receptor or the insulin receptor also takes place. This topic is also reviewed in this volume (28).

V. CONCLUDING SYNTHESIS

We have seen that equations taking account of molecular factors, such as excess molar refraction, dipolarity/polarizability, hydrogen bonding acidity and basicity, and molecular volume can closely and usefully predict passive diffusion rate of compounds of pharmaceutical interest from blood plasma into brain. Molecular structures handled by the limited number of transporters involved in facilitated diffusion and active transport at the blood–brain barrier are in some cases well characterized. Knowledge about the structures and kinetic parameters of substrates for other transporters is developing very fast and is soon likely to be complete. Once this full information has been achieved, it will be feasible to compute the permeability of smaller molecules (i.e., ≤ 2000 kDa) at the BBB by combining predictions for both passive transport and specific facilitated or active transport. Full knowledge of systemic pharmacokinetics and of molecular behavior within the brain might then lead to the ultimate goal of computation of amount and timing of doses by a given route, which will allow an optimum concentration of a drug to be maintained in the cerebral interstitial fluid.

REFERENCES

1. U Friedemann. Blood–brain barrier. Physiol Rev 22:125–245, 1942.
2. A Krogh. The active and passive exchanges of inorganic ions through the surfaces of living cells and through living membranes generally. Proc R Soc London B 133:140–200, 1946.
3. H Davson. A comparative study of the aqueous humour and cerebrospinal fluid in the rabbit. J Physiol (London) 129:11–133, 1955.

4. DP Rall, JR Stabenau, C Zubrod. Distribution of drugs between blood and cerebrospinal fluid: General methodology and effect of pH gradients. J Pharmacol Exp Ther 125:185–193, 1959.

5. SE Mayer, RP Maickel, BB Brodie. Kinetics of penetration of drugs and other foreign compounds into cerebrospinal fluid and brain. J Pharmacol Exp Ther 127:205–211, 1959.

6. SI Rapoport, H Levitan. Neurotoxicity of X-ray contrast media: Relation to lipid solubility and blood–brain barrier permeability. Am J Roentgenol 122:186–193, 1974.

7. C Hansch, AR Steward, SM Anderson, DL Bentley. The parabolic dependence of drug action upon lipophilic character as revealed by a study of hypnotics. J Med Chem 11:1–11, 1968.

8. MH Abraham. Scales of solute hydrogen bonding: Their construction and application to physicochemical and biochemical processes. Chem Soc Rev 22:73–83, 1993.

9. MH Abraham, HS Chadha, RC Mitchell. Hydrogen bonding. 33. The factors that influence the distribution of solutes between blood and brain. J Pharm Sci 83: 1257–1268, 1994.

10. JA Gratton, MH Abraham, MW Bradbury, HS Chadha. Molecular factors influencing drug transfer across the blood–brain barrier. J Pharm Pharmacol 49: 1211–1216, 1997.

11. MH Abraham, Platts JA. Physicochemical factors that influence brain uptake. In: DJ Begley, MWB Bradbury, J Kreuter, eds. The Blood–Brain Barrier and Drug Delivery to the CNS. New York: Marcel Dekker, 2000, Chap 2.

12. GW Gould, GD Holman. The glucose transporter family: Structure, function and tissue specific expression. Biochem J 295:329–341, 1993.

13. A Gjedde. Blood-brain glucose transfer. In: MWB Bradbury, ed. Handbook of Experimental Pharmacology. Vol 103, Physiology and Pharmacology of the Blood–Brain Barrier. Berlin: Springer-Verlag, 1992, pp 65–115.

14. QR Smith, J Stoll. Blood–brain barrier amino acid transport. In: WM Pardridge, ed. Introduction to the Blood–Brain Barrier. Cambridge: Cambridge University Press, 1998, pp 188–197.

15. Y Kanai, H Segawa, K Miyamoto, H Uchino, E Takeda, H Endou. Expression cloning and characterization of a transporter for large neutral amino acids activated by the heavy chain of 4F2 antigen (CD98). J Biol Chem 273:23629–23632, 1998.

16. QR Smith, M Hokari, EG Chikhale, DD Allen, PA Crooks. Computer model of the binding site of the cerebrovascular large neutral amino acid transporter. Proceedings of Cerebrovascular Biology, Salishan Lodge OR, 1998, p 212.

17. Y Takada, DT Vistica, NH Greig, I Purdon, SI Rapoport, QR Smith. Rapid high affinity transport of a chemotherapeutic amino acid across the blood–brain barrier. Cancer Res 52:2191–2196, 1992.

18. LM Albritton, L Tseng, D Scadden, JM Cunningham. A putative murine ecotropic retrovirus receptor gene encodes a multiple membrane-spanning protein and confers susceptibility to virus infection. Cell 57:659–666, 1988.

19. DA Griffith, SM Jarvis. Nucleoside and nucleobase transport systems of mammalian cells. Biochim Biophys Acta 1286:153–181, 1996.
20. EM Cornford, WH Oldendorf. Independent blood–brain barrier transport systems for nucleic acid precursors. Biochim Biophys Acta 394:211–219, 1975.
21. JR Pappenheimer, SR Heisey, EF Jourdan. Active transport of "Diodrast" and phenol sulfonaphthalein from cerebrospinal fluid to blood. Am J Physiol 200: 1–10, 1961.
22. M Bradbury. The Concept of a Blood–Brain Barrier. Chichester, UK: John Wiley, 1979, pp 197–211.
23. DJ Begley. The role of brain extracellular fluid production and effuse mechanisms in drug transport to the brain. In: DJ Begley, MWB Bradbury, J Kreuter, eds. The Blood–Brain Barrier and Drug Delivery to the CNS. New York: Marcel Dekker, 2000, Chap 6.
24. SL Wong, K van Belle, RJ Sawchuk. Distributional transport of zidovudine between plasma and brain extracellular fluid/cerebrospinal fluid in the rabbit: Investigation of the inhibitory effect of probenecid utilizing microdialysis. J Pharmacol Exp Ther 264:899–909, 1993.
25. VA Levin. Relationship of octanol/water partition coefficient and molecular weight to rat brain capillary permeability. J Med Chem 23:682–684, 1980.
26. P Borst, AH Schinkel. P-glycoprotein, a guardian of the brain. In: WM Pardridge, ed. Introduction to the Blood–Brain Barrier. Cambridge: Cambridge University Press, 1998, pp 198–206.
27. U Mayer. The relevance of P-glycoprotein in drug transport to the brain: The use of knock-out mice as a model system. In: DJ Begley, MWB Bradbury, J Kreuter, eds. The Blood–Brain Barrier and Drug Delivery to the CNS. New York: Marcel Dekker, 2000, in Chap 7.
28. U Bickel. Targetting macromolecules to the central nervous system. In: DJ Begley, MWB Bradbury, J Kreuter, eds. The Blood–Brain Barrier and Drug Delivery to the CNS. New York: Marcel Dekker, 2000, Chap 10.

2

Physicochemical Factors that Influence Brain Uptake

Michael H. Abraham and James A. Platts
University College London, London, England

I. INTRODUCTION

Numerous attempts to relate the ability of molecules to cross the blood–brain barrier (BBB) to physicochemical descriptors have been made over the last 50 years. Progress has been slow, and the tendency to use terms such as "brain penetration," "brain uptake," or "ability to cross the BBB" without exactly defining what these terms are supposed to mean has not been helpful. It is thus necessary to set out the various ways in which transfer across the blood–brain barrier has been defined (see Table 1). It is emphasized at the outset that the analyses considered in this chapter relate to passive transport across the blood–brain barrier. Specific transport mechanisms are dealt with by Akira Tsuji in Chapter 8, and Ulrich Mayer discusses the relevance of P-glycoprotein in drug transport to the brain in Chapter 7.

Biological activity was proposed as a general measure of brain uptake; in a very influential report, Hansch et al. (1) noted that the hypnotic activity of a number of congeneric series of central nervous system depressants reached a maximum when log P(oct) was near 2; P(oct) is the water–octanol partition coefficient. Later work (2) seemed to confirm this finding, and the "rule of 2" became generally accepted (3). The difficulty here is that hypnotic activity, or more generally biological activity, will depend on at least two factors: (a) a rate of transfer from blood to brain, or a distribution between blood and

Table 1 Some Measures of "Brain Uptake"

Biological activity
Maximal brain concentration
The brain uptake index from single-pass experiments
PS-product and permeability coefficient from:
Indicator dilution during single pass
Intravenous infusion or bolus injection
Vascular perfusion of brain in situ
Blood–brain distribution

brain, and (b) an interaction between drug and some receptor in the brain. If these two factors cannot be disentangled, then it is impossible to use biological activity as a measure of either rate or equilibrium transfer.

Much early work concentrated on correlations of some measure of brain uptake with log P(oct) or with some function of log P(oct). It is difficult to reach any generalization because experiments were often carried out on a set of compounds similar in chemical structure, or with too few compounds. In addition, some of the reported correlations are rather artificial. For example, in the experiments of Timmermans et al. (2), imidazoles in a series were separately injected into rats intravenously as saline solutions, and the concentration in brain, C_b, was noted at the moment of maximal decrease in blood pressure (which differed from compound to compound). The ratio C_b/C_{iv}, where C_{iv} was the initial saline concentration, was used as a measure of brain uptake. A quadratic expression relating $\log(C_b/C_{iv})$ to the water–octanol distribution coefficient, D(oct), was put forward:

$$\log(C_b/C_{iv}) = -0.94 + 0.574 \log D(\text{oct}) - 0.133 \, [\log D(\text{oct})]^2$$
$$n = 14, \, r^2 = 0.897, \, \text{SD} = 0.27, \, F = 104 \tag{1}$$

Here, and elsewhere, n is the number of data points, r is the correlation coefficient, SD is the standard deviation, and F is the F-statistic. This parabolic expression, plotted as shown in Fig. 1, gives a maximum value of $\log(C_b/C_{iv})$ at log D(oct) = 2.16, in good agreement with the rule of 2. Upon inspecting the raw data, however, it is very difficult to discern a parabola at all (see Fig. 2).

In 1998 Ducarme et al. (4) determined the enhancement of oxotremorine-induced tremors in male mice, log(PCTR), for a series of 1,3,5-triazines. As descriptors they used high performance liquid chromatography (HPLC)

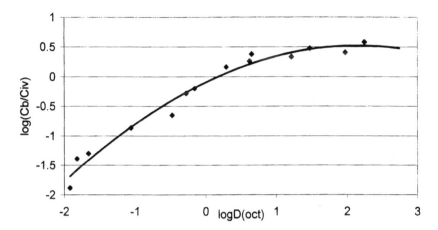

Fig. 1 Plot of $\log(C_b/C_{iv})$ against $\log D(\text{oct})$ showing the parabola of equation (1). (From Ref. 2.)

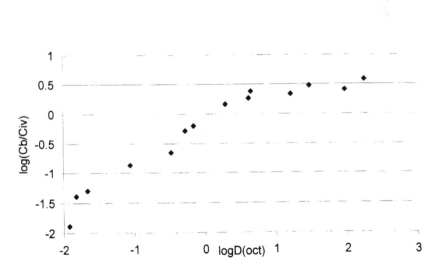

Fig. 2 Second plot of $\log(C_b/C_{iv})$ against $\log D(\text{oct})$; same data as in Fig. 1. (From Ref. 2.)

capacity factors on an immobilized artificial membrane (IAM) with an aqueous mobile phase at pH 7.4, denoted as log k'(IAM), and also log P(oct). Linear regressions were not reported, but parabolic regressions were obtained as follows:

$$\log(\text{PCTR}) = 1.880 + 1.111 \log k'(\text{IAM}) - 1.032 \, [\log k'(\text{IAM})]^2$$
$$n = 15, r^2 = 0.771, \text{SD} = 0.31, F = 20 \tag{2}$$

$$\log(\text{PCTR}) = -0.216 + 2.421 \log P(\text{oct}) - 0.635 \, [\log P(\text{oct})]^2$$
$$n = 15, r^2 = 0.851, \text{SD} = 0.25, F = 34 \tag{3}$$

The latter correlation is the better, and a much more convincing parabola is shown in Fig. 3. From equation (3), the maximum value of log(PCTR) occurs at log P(oct) = 1.9, in accord with the rule of 2. However, as we have already stated, the fact that the rule of 2 might apply to biological activity, although interesting, is not directly relevant to ability to cross the blood–brain barrier.

The brain uptake index is a more rigorous measure of brain uptake. Oldendorf (5, 6) described a method in which he obtained a relative measure of brain uptake by intracarotid injection of a mixture of ^{14}C-labeled compound and ^{3}H-labeled water, mostly from saline solution. The radioactivity in brain tissue was recorded 15 seconds after administration, and a brain uptake index (BUI) was defined as follows:

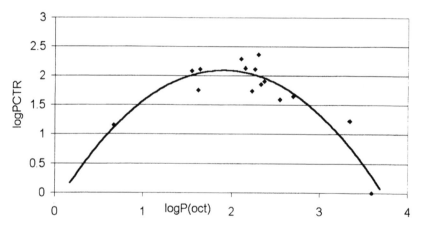

Fig. 3 Plot of log(PCTR) against log P(oct) showing the parabola of equation (3). (From Ref. 4.)

$$BUI = 100 \times \frac{(^{14}C/^3H)_{tissue}}{(^{14}C/^3H)_{saline}} \tag{4}$$

where BUI for water is 100. In additional experiments, Oldendorf showed that BUI values for amino acids were much lower when the injected solution contained rat serum; see also the work of Pardridge et al. (7) on the BUI of steroids. Although the BUI is useful as a rank order index of brain uptake, it is not easily amenable to analysis by physicochemical methods.

II. VASCULAR PERFUSION OF BRAIN

A more well-defined measure of rapid brain uptake is the permeability, expressed either as a PS product or as a permeability coefficient, obtained by Rapoport et al. (8, 9) by a method using intravenous injection and measurement of the drug profile in arterial blood. Later workers (10–12) used the in situ vascular perfusion technique, in which the rate of transfer of a radioactively labeled compound from saline or blood to brain is expressed by the same parameters, namely: a permeability–surface area product PS, in ml $s^{-1}g^{-1}$ or ml $min^{-1}g^{-1}$, or as a permeability coefficient PC, in cm $s^{-1}g^{-1}$. Both the PS product and PC are quantitative measures of the rate of transport, and so are amenable to analysis through standard physicochemical procedures. An advantage of the perfusion technique as a measure of brain uptake is that the time scale for determination of PS products is very short, as little as 15 seconds, so that back transport and biological degradation are minimized.

We have already mentioned the work of Timmermans et al. (2), which was one of the first studies to suggest a parabolic relationship with log P(oct), as noted by Hansch et al. (13)—but see Fig. 2. In a similar vein, physicochemical analyses of brain perfusion have mostly used either log PC or log PS in various correlations with function of log P(oct). Ohno et al. (8) showed that log$[PC \cdot MW^{1/2}]$ was proportional to log P(oct), for three compounds only; MW is the compound molecular weight. Rapoport et al. (9) showed that there was a linear relationship between log PC and log P(oct) for 17 compounds but gave no numerical data. In contrast to these studies, Pardridge et al. (12) plotted their in vivo PS products, as log$[PS \cdot MW^{1/2}]$, against log P(oct), but only a poor correlation was obtained, with $r^2 = 0.72$; actually, incorporation of the factor $MW^{1/2}$ makes little difference.

More recently, there have been a number of studies of permeation in which novel descriptors have been used. Kai et al. (14) have correlated the old (graphical) data of Rapoport et al. (9) with a parameter P_H, obtained from the temperature variation of water–micelle partition coefficients,

$$\log PC = -5.12 + 0.33 \, P_H$$
$$n = 8, \, r^2 = 0.90, \, SD = 0.41, \, F = 53 \qquad (5)$$

Only 8 of the 17 compounds shown by Rapoport et al. (9) were studied, and it remains to be seen how useful is this correlation.

A much more useful study is that of Burton et al. (15), who set up an ethylene glycol–heptane partitioning system, $P(eh)$, as a model for permeability across *Caco-2* cells (16) and for permeability from saline to rat brain (17, 18). In the latter case they showed that for seven peptides log PS was well correlated with log $P(eh)$ or with Δ log P, but poorly correlated with log $P(oct)$. Correlation coefficients were 0.940 against log $P(eh)$, 0.962 against Δ log P, and 0.155 against log $P(oct)$ for uncorrected *PS* values (17), and 0.970, 0.874, and 0.598, respectively, for values corrected for efflux (18). The parameter Δ log P was defined by Seiler (19) as follows:

$$\Delta \log P = \log P(oct) - \log P(cyc) \qquad (6)$$

where $P(cyc)$ is the water–cyclohexane partition coefficient.

Quite recently, Gratton et al. (20, 21) have determined *PS* products by vascular perfusion from saline to brain in rats for 18 very varied compounds. They found reasonable correlations of log PS with log $P(oct)$, equations (7) and (8), although it is doubtful that any parabolic relationship exists (see Fig. 4.)

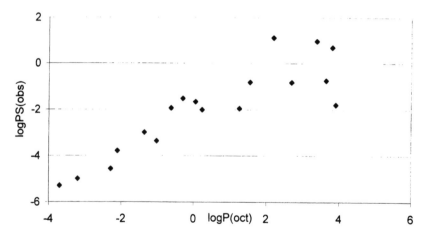

Fig. 4 Plot of log *PS* against log $P(oct)$. (From Ref. 21).

$$\log PS = -2.28 + 0.69 \log P(\text{oct})$$
$$n = 18, \ r^2 = 0.777, \ SD = 0.94, \ F = 55 \tag{7}$$

$$\log PS = -1.81 + 0.75 \log P(\text{oct}) - 0.084 \log P \ (\text{oct})]^2$$
$$n = 18, \ r^2 = 0.829, \ SD = 0.85, \ F = 36 \tag{8}$$

It would have been useful to correlate log PS with the $\Delta \log P$ parameter (19), but many values of log $P(\text{cyc})$ were unavailable. This is an oft-encountered difficulty in the use of the $\Delta \log P$ parameter.

Although there are numerous physicochemical studies on brain perfusion, it is not possible to reach any general conclusions. This is due to a number of factors. First, different workers use slightly different experimental techniques and, most importantly, different perfusate solutions. Perfusion from saline will not be the same as perfusion from saline containing albumin, and neither will be the same as perfusion from blood. Second, as we have noted, many studies involve too few compounds to permit any generalization to be made. Third, even if a reasonable number of compounds are used, any generalization will be difficult if the compounds are all of the same type.

III. BLOOD–BRAIN DISTRIBUTION

The equilibrium distribution of compounds between blood and brain is a much longer-term measure of brain uptake than is brain perfusion, with experiments that can last up to several days. The work of Young and Mitchell and their colleagues (22, 23) marked a decisive step forward, and nearly all physico-analyses of blood–brain distribution have used the Young–Mitchell (YM) data set for in vivo distribution ratios in rats, defined as follows:

$$BB = \frac{\text{conc. in brain}}{\text{conc. in blood}} \tag{9}$$

In their extensive study on blood–brain distribution, these workers (22) determined log $P(\text{cyc})$ as well as log $P(\text{oct})$ for many of their drug compounds, and so were able to test $\Delta \log P$ and log $P(\text{oct})$ as possible predictors of log BB. For 20 compounds they found for log $P(\text{oct})$ that $r^2 = 0.190$, SD = 0.71, and $F = 4$; for log $P(\text{cyc})$, $r^2 = 0.536$, SD = 0.54, and $F = 21$; for $\Delta \log P$, $r^2 = 0.690$, SD = 0.44, and $F = 40$. Hence log $P(\text{oct})$ is quite useless as a predictor of log BB, but $\Delta \log P$ is very much better. Sometime later, these correlations were extended by the inclusion of all the YM log BB values (not just the 20 considered before) and by addition of compounds studied by indirect in vitro methods (24, 25). The most recent correlations (26) are as follows:

$$\log BB = -0.38 + 0.23 \log P(\text{oct})$$
$$n = 51, r^2 = 0.296, \text{SD} = 0.50, F = 21 \tag{10}$$

$$\log BB = 0.35 - 0.28 \, \Delta \log P$$
$$n = 48, r^2 = 0.650, \text{SD} = 0.32, F = 85 \tag{11}$$

thus confirming the original conclusions (22) on the use of log $P(\text{oct})$ and $\Delta \log P$ as predictors of log BB. A plot of log BB against log $P(\text{oct})$ for the 51 compounds is shown in Fig. 5. The scatter is not random, and compounds that are acids with strong hydrogen bonds (shown as diamonds) have observed log BB values much lower than predicted. This is not so for the corresponding plot against $\Delta \log P$ (see Fig. 6), where the scatter is largely random.

Kaliszan et al. (27) attempted to improve on the poor correlation with log $P(\text{oct})$ by including MW, the compound molecular weight, as an additional descriptor. For the restricted set of 20 YM compounds, however, only a moderately useful equation was obtained:

$$\log BB = 0.476 + 0.541 \log P(\text{oct}) - 0.00794 \, \text{MW}$$
$$n = 20, r^2 = 0.642, \text{SD} = 0.49, F = 15 \tag{12}$$

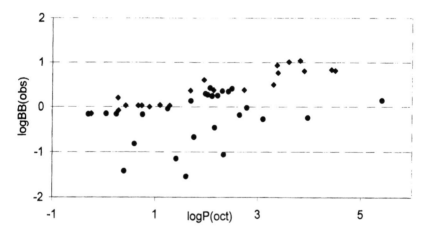

Fig. 5 Plot of log BB against log $P(\text{oct})$; diamonds, nonacids; circles, hydrogen bond acids. (From Ref. 26.)

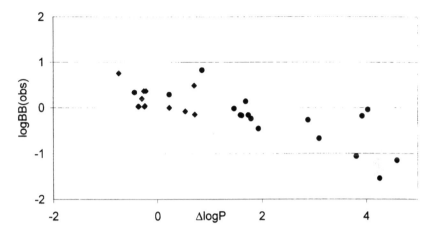

Fig. 6 Plot of log BB against Δ log P; diamonds, nonacids; circles, hydrogen bond acids. (From Ref. 26.)

although the corresponding equation using log P(cyc) gave better results. For 33 compounds of the extended data set of Abraham et al. (24), equation (13) was obtained, although why the method was restricted to only part of the extended set is not clear:

$$\log BB = -0.088 + 0.272 \log P(\text{oct}) - 0.00116 \text{ MW}$$
$$n = 33, r^2 = 0.897, \text{SD} = 0.13, F = 131 \tag{13}$$

A "size" correction to log P(oct) was also used by Salminen et al. (28), who studied a data set quite different from that commonly used (22, 24) and obtained the equation

$$\log BB = 1.25 + 0.35 \log P(\text{oct}) - 0.01 V_m + 0.99 I_3$$
$$n = 23, r^2 = 0.848, \text{SD} = 0.32, F = 35 \tag{14}$$

where V_m is a calculated van der Waals solute volume and I_3 is an indicator variable that is zero except for compounds with an amino nitrogen ($+1$) or with a carboxyl group (-1). Three compounds were removed as outliers to obtain equation (14). Similar statistics were obtained if log k'(IAM) was used as a descriptor (cf. Ref. 4):

$$\log BB = 1.28 + 0.58 \log k'(\text{IAM}) - 0.01 V_m + 0.89 I_2$$
$$n = 21, r^2 = 0.848, \text{SD} = 0.27, F = 31 \tag{15}$$

The I_2 descriptor is zero except for compounds with amino nitrogen $(+1)$; five compounds were omitted to obtain equation (15).

Van de Waterbeemd and Kansy (29) also analyzed 20 compounds in the YM data set and put forward two equations for the correlation of log BB values:

$$\log BB = 1.730 - 0.338 \, \Lambda(\text{alk}) + 0.007 \, V_M$$
$$n = 20, \, r^2 = 0.872, \, \text{SD} = 0.29, \, F = 58 \tag{16}$$

$$\log BB = 1.643 - 0.021 \, SP - 0.003 \, V_M$$
$$n = 20, \, r^2 = 0.697, \, \text{SD} = 0.45, \, F = 20 \tag{17}$$

In these equations, $\Lambda(\text{alk})$ is found from experimental water–alkane partition coefficients as the difference in log $P(\text{alk})$ between the experimental value and that calculated from a reference line for alkanes. Note that in practice, water–cyclohexane partition coefficients were actually used (29). As before, V_M is a calculated van der Waals solute volume, and SP is the calculated hydrophilic part of the van der Waals surface:

$$\log P(\text{alk}) = a \, V_M + \Lambda(\text{alk})$$
$$n = 8, \, r^2 = 1.000, \, \text{SD} = 0.04, \, F = 11830 \tag{18}$$

Abraham et al. (24) pointed out that the large intercepts in equations (16) and (17) meant that log BB values for smaller sized compounds would be incorrectly predicted, and Calder and Ganellin (30) showed that equation (17) greatly overestimated log BB values for a variety of compounds.

Thus of the correlations, equations (10) to (17), the only equation that both is general enough and has a reasonably good statistical fit is equation (11). In addition, none of these equations lead to any easy interpretation of the physicochemical factors that influence blood–brain distribution. To achieve such interpretation, we set out in the next section the development of correlation equations that include quantitative hydrogen bond descriptors.

IV. CORRELATION EQUATIONS USING HYDROGEN BOND DESCRIPTORS

Although measures of proton acidity and proton basicity have been part of the fabric of chemistry for the past 100 years, only recently have scales of acidity and basicity been developed for hydrogen bonds. In the present context, the most useful hydrogen bond scales are those that measure the "overall" or "effective" hydrogen bond acidity and basicity when a solute is surrounded by solvent molecules. These scales are denoted as $\Sigma\alpha_2^H$ and $\Sigma\beta_2^H$ (31), and numerous tables of hydrogen bond acidity and basicity have been published (31–34). A selection of values of $\Sigma\alpha_2^H$ and $\Sigma\beta_2^H$ is given in Table 2. Across families of compounds there is little connection with proton acidity or basicity. Acetic acid and phenol have almost the same $\Sigma\alpha_2^H$ value, but their pK_a values in water differ by 5 log units. Likewise, acetamide is a very weak proton base, but it has almost the same $\Sigma\beta_2^H$ value as trimethylamine, a strong proton base. It is also clear that the practice of using an indicator variable for hydrogen bond acidity or basicity is a very approximate procedure. Table 3 gives $\Sigma\alpha_2^H$ values for a number of solutes that have either one N—H bond or one O—H bond. The spread of $\Sigma\alpha_2^H$ values is very large, so quantitative scales are required to ensure the proper inclusion of hydrogen bond effects of solutes.

Once scales of hydrogen bonding were set out, it was possible to use them as solute descriptors in a general equation for transport properties. These can be defined as properties in which the main process is either the equilibrium

Table 2 Values of the Hydrogen Bond Descriptors $\Sigma\alpha_2^H$ and $\Sigma\beta_2^H$

Solute	$\Sigma\alpha_2^H$	$\Sigma\beta_2^H$
Hexane	0.00	0.00
Ethyne	0.06	0.04
Trichloromethane	0.15	0.02
Butanone	0.00	0.51
Acetonitrile	0.04	0.33
Acetamide	0.54	0.68
Dimethylsulfone	0.00	0.76
Diethylamine	0.08	0.68
Trimethylamine	0.00	0.67
Acetic acid	0.61	0.45
Benzene	0.00	0.14
Chlorobenzene	0.00	0.07
Benzenesulfonamide	0.55	0.80
Pyridine	0.00	0.52
Phenol	0.60	0.30

Source: Refs. 26, 31–34.

Table 3 Values of $\Sigma\alpha_2^H$ for
Some Monacidic N—H and O—H
Acids

Solute	$\Sigma\alpha_2^H$
t-Butanol	0.31
Methanol	0.43
2,2,2-Trifluoroethanol	0.57
Phenol	0.60
Acetic acid	0.61
Hexafluoropropan-2-ol	0.77
4-Nitrophenol	0.82
Trichloroacetic acid	0.95
Diethylamine	0.08
Aniline	0.26
N-Methylacetamide	0.40
Indole	0.44
Pyrazole	0.54

Source: Refs. 26, 31–34.

transfer of a solute from one phase to another or a rate of transfer from one phase to another. The general equation takes the form,

$$\log SP = c + r\,R_2 + s\,\pi_2^H + a\,\Sigma\alpha_2^H + b\,\Sigma\beta_2^H + v V_x \tag{19}$$

where *SP* is a set of solute properties in a given system. For example, SP can be a set of *BB* values for a series of solutes (31). The independent variables are solute descriptors as follows: R_2 is an excess molar refraction, π_2^H is the dipolarity/polarizability, $\Sigma\alpha_2^H$ and $\Sigma\beta_2^H$ are the hydrogen bond descriptors, and V_x is the solute McGowan volume (35) in units of cubic centimeters per mole divided by 100. Values of R_2 and V_x can be calculated from structure, and the remaining three descriptors, π_2^H, $\Sigma\alpha_2^H$, and $\Sigma\beta_2^H$, are obtained from experimental measurements such as log *P* values and HPLC log *k'* values, as described in detail elsewhere (36).

Because equation (19) was constructed using descriptors that refer to specific solute physicochemical properties, the coefficients encode properties of the system. This can be shown by considering the correlation equation for the solubility of gases and vapors in water, with $SP = L^w$, the gas–water partition coefficient defined through L^w = (conc of solute in the gas phase)/(conc of solute in water).

Application of equation (19) to a series of log L^w values gave (34) the following equation:

$$\log L^w = -0.994 + 0.577R_2 + 2.549\pi_2^H$$
$$+ 3.813\Sigma\alpha_2^H + 4.841\Sigma\beta_2^H - 0.869V_x \qquad (20)$$
$$n = 408, \; r^2 = 0.995, \; SD = 0.15, \; F = 16810$$

The positive r coefficient shows that the solute molecules interact with the surrounding solvent through σ- and π-electron pairs, the s coefficient relates to the solvent dipolarity/polarizability, the a coefficient relates to the solvent hydrogen bond basicity (because acidic solutes will interact with basic solvents), and the b coefficient to the solvent hydrogen bond acidity. The v coefficient is the resultant of an exoergic solute–solvent general dispersion interaction (that leads to a positive coefficient) and an endoergic solvent cavity effect (that leads to a negative coefficient). The first four terms in equation (20) are all positive, indicating exoergic solute–solvent interactions, and the negative v coefficient shows that for water as solvent, the endoergic cavity effect is larger than the exothermic solute–water dispersion interaction.

For partition between two phases, the coefficients in equation (19) will refer to the difference in properties of the two phases. A very important partition is that between water and octanol, for which the correlation equation (34) is

$$\log P(\text{oct}) = 0.088 + 0.562R_2 - 1.054\pi_2^H$$
$$+ 0.034\Sigma\alpha_2^H - 3.460\Sigma\beta_2^H + 3.814V_x \qquad (21)$$
$$n = 613, \; r^2 = 0.995, \; SD = 0.12, \; F = 23162$$

The coefficients in equation (21) show that water is more dipolar than (wet) octanol (because the s coefficient is negative) and more acidic than octanol (because the b coefficient is negative). As expected, the v coefficient is positive, so that increase in volume leads to an increase in log $P(\text{oct})$. What is unexpected is that the a coefficient is nearly zero, so that solute acidity has no effect on log $P(\text{oct})$; we shall see that this is a very important observation. As a corollary, it can be deduced that water and octanol have the same hydrogen bond basicity.

Another descriptor often used to correlate log BB values is the Seiler Δ log P parameter, defined (19) through equation (6) and sometimes regarded as a measure of solute hydrogen bond acidity (37). Application of equation (19) shows that solute hydrogen bond acidity is certainly a major term but that other terms, especially hydrogen bond basicity, are important (34):

$$\Delta \log P(16) = -0.072 - 0.093R_2 + 0.528\pi_2^H$$
$$+ 3.655\Sigma\alpha_2^H + 1.396\Sigma\beta_2^H - 0.521V_x \qquad (22)$$
$$n = 288, \ r^2 = 0.967, \ SD = 0.17, \ F = 1646$$

Equation (22) actually refers to hexadecane as the alkane phase, rather than to cyclohexane, but the difference is very small (34).

V. HYDROGEN BONDING AND BRAIN PERFUSION

The general equation (19) has been applied to the *PS* products for vascular perfusion from saline to rat brain by Gratton et al. (21), who obtained the correlation equation

$$\log PS = -1.21 + 0.77R_2 - 1.87\pi_2^H$$
$$- 2.80\Sigma\beta_2^H + 3.31V_x \qquad (23)$$
$$n = 18, \ r^2 = 0.953, \ SD = 0.48, \ F = 65$$

The 18 compounds ranged from sucrose (log *PS* = −5.30) to propanolol (log *PS* = 0.98) with *PS* in units of ml s^{-1}g^{-1}. All the *PS* products were corrected to refer to compounds in their neutral form. Equation (23) is not particularly good, but the SD value of 0.48 log unit is much smaller than those in equations (7) and (8). Figure 7 plots observed versus calculated log *PS* values, showing random scatter about the line of identity (cf. Fig. 4).

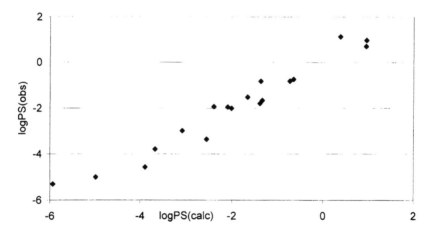

Fig. 7 Plot of log *PS* observed against log *PS* calculated on equation (23). (From Ref. 21.)

Furthermore, equation (23) shows exactly the factors that influence log *PS*. Interactions with σ- and π-electron pairs, as well as solute size, increase log *PS*; and solute dipolarity/polarizability and hydrogen bond basicity reduce log *PS*. Very interestingly, solute hydrogen bond acidity has no effect on log *PS*, which is why log *P*(oct) is a reasonable descriptor for log *PS* [see equation (7) and Fig. 4].

The factors influencing log *PS* can be put on a quantitative basis through a term-by-term analysis of equation (23), as shown in Table 4. The two largest terms are $b \cdot \Sigma\beta_2^H$, which is always negative and leads to small *PS* products, and $v \cdot V_x$, which is always positive and leads to large *PS* products. The term $a \cdot \Sigma\alpha_2^H$ is always zero and so has been omitted from equation (23).

Chikhale et al. (17), as described above, correlated log *PS* values for perfusion of seven peptides into rat brain, with log *P*(eh) for the novel partitioning system ethylene glycol–heptane. The success of the correlation, $r^2 = 0.884$, led them to suggest that ''hydrogen bond potential'' was a major factor influencing transport of these peptides, with log *P*(eh) as a measure of hydrogen bond potential. Using the hydrogen bond approach of Abraham, it is possible to be more specific. Paterson et al. (16) had already correlated log *P*(eh) with various descriptors, including hydrogen bond descriptors, for 19 compounds; but Abraham et al. (38) were able to obtain a much more general relationship for 75 compounds:

$$\log P(\text{eh}) = 0.336 - 0.075R_2 - 1.201\pi_2^H$$
$$- 3.786\Sigma\alpha_2^H - 2.201\Sigma\beta_2^H + 2.085V_x \qquad (24)$$
$$n = 75, \; r^2 = 0.966, \; \text{SD} = 0.28, \; F = 386$$

Table 4 Term-by-Term Analysis of Equation (23) for log PS Values[a]

Solute	r (R_2)	s (π_2^H)	b $(\Sigma\beta_2^H)$	v (V_x)	log PS$_{calc}$	log PS$_{obs}$
Sucrose	1.52	−4.68	−8.96	7.37	−5.96	−5.30
Mannitol	0.64	−3.37	−5.38	4.32	−5.00	−5.01
Urea	0.37	−1.87	−2.52	1.54	−3.69	−3.79
Thymine	0.62	−1.86	−2.88	2.95	−2.40	−1.93
Ethanol	0.19	−0.80	−1.34	1.49	−1.67	−1.52
Propanolol	1.42	−2.81	−3.56	7.11	0.95	0.98

[a] There is a constant of −1.21 in each row.
Source: Ref. 21.

Hydrogen bond descriptors were obtained (38) for the peptides studied by Chikhale et al. (17), and the factors involved in the partition of the peptides between ethylene glycol and heptane were evaluated. This can be achieved only if both the coefficients in the process and the solute descriptors are known, as shown in Table 4. Hence until the perfusion system of Chikhale et al. (17) has been characterized [cf. equation (23)], the influence of peptide hydrogen bond acidity or basicity on the respective log *PS* values cannot be fully understood.

VI. HYDROGEN BONDING AND BLOOD–BRAIN DISTRIBUTION

The first correlation of blood–brain distribution coefficients (as log *BB*) using hydrogen bond descriptors was that of Abraham et al. (24, 25, 39, 40). The in vivo values of log *BB* were available (28) for 30 solutes (not just the 20 YM values that were invariably selected), and a number of indirect in vitro values were obtained from Abraham and Weathersby (41), giving a total of 65 log *BB* values. Abraham et al. (24) first showed that application of equation (19) to the in vitro data set gave a correlation equation very similar to that for the in vivo data set. This is an important observation because it means that the two data sets are compatible and can be combined to give a very varied data set of log *BB* values. Application of equation (19) to the combined set gave (24) the regression equation

$$\log BB = -0.038 + 0.198R_2 - 0.687\pi_2^H$$
$$- 0.715\Sigma\alpha_2^H - 0.698\Sigma\beta_2^H + 0.995V_x \qquad (25)$$
$$n = 57, \ r^2 = 0.907, \ SD = 0.20, \ F = 99$$

This equation is markedly better than the general equations (10) and (11); compare Fig. 8 with Figs. 5 and 6. In addition, equation (25) is of considerable interest because it shows, for the first time, exactly the solute factors that influence log *BB*. General dispersion interactions (R_2) and solute size (V_x) favor brain; and solute hydrogen bond acidity, hydrogen bond basicity, and dipolarity/polarizability favor blood.

Eight solutes were considerable outliers to equation (25); although they were not randomly scattered about the line of identity of equation (25), all had observed values of log *BB* much less than calculated. This could be due to two effects. First, several hours may have to elapse before equilibrium is

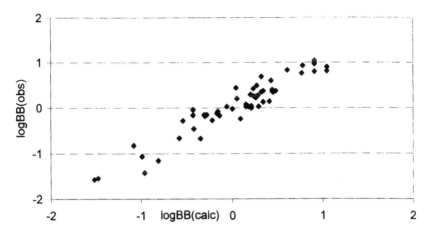

Fig. 8 Plot of log *BB* observed against log *BB* calculated on equation (25). (From Ref. 24.)

established in the distribution experiments, and biological degradation of some of the compounds is to be expected. If compounds are broken down into smaller fragments (i.e., with smaller V_x values) that are retained more in blood, then the radioassay method used will lead to smaller observed log *BB* values. Second, it is now known (18, 42) that compounds may be transported out of the BBB by an efflux pump mechanism involving P-glycoprotein. Operation of such a mechanism will lead to smaller observed log *BB* values than were expected. There are therefore considerable biochemical reasons for omitting compounds that all have much lower log *BB* values than calculated.

The coefficients in equation (25) show also why log *P*(oct) is a poor descriptor for log *BB*. In equation (25), the term in $\Sigma\alpha_2^H$ is just as large as the term in $\Sigma\beta_2^H$; but in the corresponding equation (21) for log *P*(oct), there is no term in $\Sigma\alpha_2^H$ at all. Hence compounds that are hydrogen bond acids will be out of line on equation (21), exactly as shown in Fig. 5. However, the Δ log *P* equation (22) contains terms in π_2^H, $\Sigma\alpha_2^H$, $\Sigma\beta_2^H$, and V_x, all of which have a substantial influence, and with the correct relative signs. Hence there is no particular effect of acids in disrupting the correlation, but only a general scatter (see Fig. 6).

Values of log *BB* for a few compounds not used in the log *BB* equation were later predicted, and the predictions shown (40) to be in agreement with experiment (23) using equation (25). Predictions using log *P*(oct) or Δ log *P* were rather poor (40).

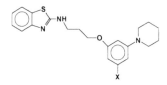

Fig. 9 Zolantidine (X = H).

Chadha et al. (25) showed how easy it was to apply equation (25) by considering the case of substituted zolantidines (see Fig. 9). The effect of a group (X) on log *BB* can simply be predicted through the estimation of descriptors. In effect, the descriptors compared to X = H are just aromatic substituents. Details are in Table 5, which shows how log *BB* can be altered at will, by choice of substituents. It should be noted that this is not an example of a Hammett-type equation, because all the substituent effects are predicted from structure.

Two recent calculations of log *BB* may usefully be compared to the method above. Lombardo et al. (43) have correlated log *BB* with a calculated Gibbs energy of transfer of compounds from the gas phase to water, ΔG_{w}^{0}. Equation (26) was obtained, with ΔG_{w}^{0} in kilocalories per mole:

$$\log BB = 0.43 + 0.054 \, \Delta G_{w}^{0}$$
$$n = 55, \; r^2 = 0.671, \; SD = 0.41, \; F = 108 \tag{26}$$

Since the 55-compound set of Lombardo et al. (43) was largely based on the full set studied by Abraham et al. (24), equation (26) can be compared

Table 5 Prediction of Substituent Effects in Zolantidine on log *BB*

X	R_2	π_2^H	$\Sigma\alpha_2^H$	$\Sigma\beta_2^H$	V_x	log *BB*
H	2.69	2.64	0.40	1.38	2.9946	0.42
Me	2.69	2.67	0.40	1.39	3.1355	0.53
Et	2.69	2.67	0.40	1.39	3.2764	0.67
Ph	3.44	3.11	0.40	1.50	3.6134	0.77
Cl	2.81	2.75	0.40	1.31	3.1170	0.54
OMe	2.80	2.90	0.40	1.54	3.1942	0.35
OH	2.82	3.06	0.99	1.48	3.0533	−0.28
NH$_2$	3.01	3.11	0.65	1.64	3.0944	−0.11

directly with equation (23), compare also Fig. 10 with Fig. 8. The success of equation (26), the correlation equation, seems to derive from the very small slope, so that large errors in the calculated ΔG_w^0 value are attenuated into small errors in the calculated log BB value. However, the correlation is clearly useful in that values of log BB can be calculated from structure, using no experimental data at all. Six compounds were used as a test set; predicted and observed log BB values agreed, with SD = 0.62 log unit, in line with the SD value for equation (26).

The transfer of a solute from the gas phase to water at first sight seems to be an unlikely model for transfer from blood to brain. However, we can analyze the two processes through the method of Abraham. Values of ΔG_w^0 are proportional to $-\log L^w$, an equation for which is given as equation (21). Now for two processes to be matched to each other, it is not necessary for the two sets of coefficients in equation (19) to be matched, but only their ratios. If the coefficients in equation (21) are divided by -6, the ratios in Table 6 are obtained, and these can be compared with coefficients in the log BB equation. There is very good agreement for the three polar terms: the v coefficient has the correct sign, and only the not-very-important r coefficient has the wrong sign. So, perhaps unexpectedly, the gas–water transfer does yield ratios of coefficients not far from those for log BB.

The most recent work on correlations of log BB is that of Norinder et al. (44), who calculated 14 descriptors including log P(oct), hydrogen bond

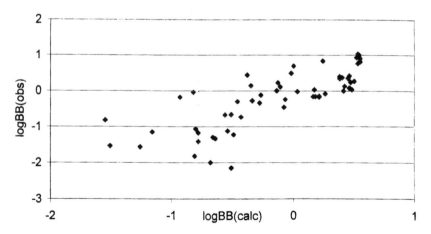

Fig. 10 Plot of log BB observed against log BB calculated on equation (26). (From Ref. 43.)

Table 6 Comparison of Coefficients for Gas–Water and Blood–Brain Transfers

Equation	r	s	a	b	v
log L^u	0.577	2.549	3.813	4.841	−0.869
$(-\log L^u)/6$	−0.096	−0.425	−0.635	−0.807	0.145
log BB	0.198	−0.687	−0.715	−0.698	0.995

acidity, hydrogen bond basicity, Lewis acidity, and Lewis basicity. These 14 calculated descriptors were combined into three significant partial least squares (PLS) components that were used as descriptors for the correlation of log BB values. The latter were exactly those used by Lombardo et al. (43), hence were nearly the same as those used by Abraham et al. (24).

Norinder et al. (44) set up two models. In the first model, 28 log BB values were correlated to the three PLS components with $r^2 = 0.862$, SD = 0.31, and $F = 50$; values of the remaining 28 log BB values were then predicted, with SD = 0.35 log unit. The second model incorporated all 56 log BB values, which were correlated to three PLS components with $r^2 = 0.834$, SD = 0.31, and $F = 87$; observed and calculated log BB values are plotted in Fig. 11. Note that the PLS components in model 2 contained proportions of the 14 descriptors different from those in model 1. Also the F-statistic is somewhat misleading because only three independent variables are used in the correlations, but 14 descriptors are actually used. Nevertheless, the SD

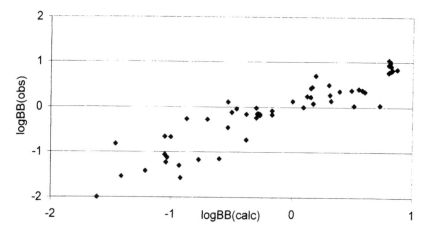

Fig. 11 Plot of log BB observed against log BB calculated on model 2. (From Ref. 44.)

Table 7 Correlations of log *BB* for Extended Data Sets

Ref.	n	r^2	SD	F	Descriptors
26, eq. 10	51	0.296	0.50	21	log *P*(oct)
26, eq. 11	48	0.650	0.32	85	Δ log *P*
24, eq. 25	57	0.907	0.20	99	Five
43, eq. 26	55	0.671	0.41	108	ΔG_w^0
44, model 1	28	0.862	0.31	50	Three (fourteen)
44, model 2	56	0.834	0.31	87	Three (fourteen)

values of 0.31–0.35 for log *BB* represent the best agreement between experimental and calculated log *BB* values for any computation from structure yet reported.

We summarize in Table 7 the various correlations of log *BB* values for reasonably extended data sets (see also Figs. 5, 6, 8, 10, and 11). The log *P*(oct) descriptor is of little use in any general context, but the Δ log *P* descriptor leads to a useful statistical correlation. Unfortunately, the experimental determination of water–cyclohexane partition coefficients needed to calculate Δ log *P* is often difficult. The two most useful correlations are those of Abraham et al. (24) and Norinder et al. (44). The former is statistically good and has the advantage of being easily interpreted in general chemical terms, while the latter has an advantage that log *BB* values can be calculated from structure. Platts et al. (45) have just worked out a computerized procedure to calculate the Abraham descriptors from structure, which would then allow the calculation of log *BB* values from structure. It remains to be seen how well this method compares with the calculational methods of Lombardo et al. (43) and Norinder et al. (44).

Finally, we note that correlations for log *BB* are not very similar to correlations for log *PS*. This can be seen from Figs. 1–5, the various plots of log *BB* or log *PS* against log *P*(oct). Hence the factors that influence blood–brain distribution are not quantitatively the same as those that influence brain perfusion. As we stressed at the beginning of this chapter, it is vitally important when discussing "brain uptake" to specify what measure of brain uptake is being used.

ACKNOWLEDGMENTS

We are grateful to Glaxo-Wellcome for a postdoctoral fellowship (to JAP), and we thank Michael W. B. Bradbury for his help and interest.

REFERENCES

1. C Hansch, AR Steward, SM Anderson, DL Bentley. The parabolic dependence of drug action upon lipophilic character as revealed by a study of drug action. J Med Chem 11:1–11, 1968.

2. PBMWM Timmermans, A Brands, PA van Zwieten. Lipophilicity and brain disposition of clonidine and structurally related imidazolidines. Arch Pharmacol 300:217–226, 1977.

3. SP Gupta. QSAR studies on drugs acting at the central nervous system. Chem Rev 89:1765–1800, 1989.

4. A Ducarme, M Neuwels, S Goldstein, R Massingham. IAM retention and blood brain barrier penetration. Eur J Med Chem 33:215–223, 1998.

5. WH Oldendorf. Measurement of brain uptake of radiolabeled substances using a tritiated water internal standard. Brain Res 24:372–376, 1970.

6. WH Oldendorf. Brain uptake of radiolabeled amino acids, amines, and hexoses after arterial injection. Am J Physiol 221:1629–1639, 1971.

7. WM Pardridge, LJ Mietus. Transport of steroid hormones through the rat blood–brain barrier. J Clin Invest 54:145–154, 1979.

8. K Ohno, KD Pettigrew, SI Rapoport. Lower limits of cerebrovascular permeability to nonelectrolytes in the conscious rat. Am J Physiol 235:H299–H307, 1978.

9. SI Rapoport, K Ohno, KD Pettigrew. Drug entry into the brain. Brain Res 172: 354–359, 1979.

10. Y Takasato, SI Rapoport, QR Smith. An in situ brain perfusion technique to study cerebrovascular transport in the rat. Am J Phys 247:H484–H493, 1984.

11. R Deane, MWB Bradbury. Transport of lead-203 at the blood–brain barrier during short cerebrovascular perfusion with saline in the rat. J Neurochem 54:905–914, 1990.

12. WM Pardridge, D Triguero, J Yang, PA Cancilla. Comparison of in vitro and in vivo models of drug transcytosis through the blood–brain barrier. J Pharm Exp Ther 253:884–891, 1990.

13. C Hansch, JP Bjorkroth, A Leo. Hydrophobicity and central nervous system agents: On the principle of minimum hydrophobicity in drug design. J Pharm Sci 76:663–687, 1987.

14. J Kai, K Nakamura, T Masuda, I Ueda, H Fujiwara. Thermodynamic aspects of hydrophobicity and the blood–brain barrier permeability studied with a gel filtration chromatography. J Med Chem 39:2621–2624, 1996.

15. PS Burton, RA Conradi, AR Hilgers, NFH Ho, LL Maggiora. The relationship between peptide structure and transport across epithelial cell monolayers. J Controlled Release 19:87–98, 1992.

16. DA Paterson, RA Conradi, AR Hilgers, TI Vidmar, PS Burton. A non-aqueous partitioning system for predicting the oral absorption potential of peptides. Quant Struct-Act Relat 13:4–10, 1994.

17. EG Chikhale, K-Y Ng, PS Burton, RA Borchardt. Hydrogen bonding potential

as a determinant of the in vitro and in situ blood–brain permeability of peptides. Pharm Res 11:412–419, 1994.

18. EG Chikhale, PS Burton, RA Borchardt. The effect of verapamil on the transport of peptides across the blood–brain barrier in rats: Kinetic evidence for an apically polarized efflux mechanism. J Pharmacol Exp Ther 273:298–303, 1995.

19. P Seiler. Interconversion of lipophilicities from hydrocarbon/water systems into octanol/water systems. Eur J Med Chem 9:473–479, 1974.

20. JA Gratton, SL Lightman, MW Bradbury. Transport into retina measured by short vascular perfusion in the rat. J Physiol 470:651–663, 1993.

21. JA Gratton, MH Abraham, MW Bradbury, HS Chadha. Molecular factors influencing drug transport across the blood–brain barrier. J Pharm Pharmacol 49: 1211–1216, 1997.

22. RC Young, RC Mitchell, TH Brown, CR Ganellin, R Griffiths, M Jones, KK Rana, D Saunders, IR Smith, NE Sore, TJ Wilks. Development of a new physicochemical model for brain penetration and its application to the design of centrally acting H2 receptor histamine antagonists. J Med Chem 31:656–671, 1988.

23. RC Young, CR Ganellin, R Griffiths, RC Mitchell, ME Parsons, D Saunders, NE Sore. An approach to the design of penetrating histaminergic agonists. Eur J Med Chem 28:201–211, 1993.

24. MH Abraham, HS Chadha, RC Mitchell. Hydrogen bonding. 33. Factors that influence the distribution of solutes between blood and brain. J Pharm Sci 83: 1257–1268, 1994.

25. HS Chadha, MH Abraham, RC Mitchell. Physicochemical analysis of the factors governing distribution of solutes between blood and brain. Bioorg Med Chem Lett 4:2511–2516, 1994.

26. MH Abraham, unpublished work.

27. R Kaliszan, M Markuszewski. Brain/blood distribution described by a combination of partition coefficient and molecular mass. Int J Pharm 145:9–16, 1996.

28. T Salminen, A Pulli, J Taskinen. Relationship between artificial immobilised membrane chromatographic retention and the brain penetration of structurally diverse drugs. J Pharm Biomed Anal 15:469–477 (1997).

29. H van de Waterbeemd, M Kansy. Hydrogen-bonding capacity and brain penetration. Chemia 46:299–303, 1992.

30. JAD Calder, CR Ganellin. Predicting the brain-penetrating capability of histaminic compounds. Drug Design Discuss 11:259–268, 1994.

31. MH Abraham. Scales of solute hydrogen-bonding: Their construction and application to physicochemical and biochemical processes. Chem Soc Rev 22:73–83, 1993.

32. MH Abraham, HS Chadha. Application of a solvation equation to drug transport properties. In: V Pliska, H van de Waterbeemd, B Testa, eds. Lipophilicity in Drug Action and Toxicology. VCH, Weinheim, Germany, 1996, pp 311–337.

33. MH Abraham, J Andonian-Haftvan, GS Whiting, A Leo, RW Taft. Hydrogen bonding. 34. The factors that influence the solubility of gases and vapours in

water at 298 K, and a new method for its determination. J Chem Soc Perkin Trans 2 1994, 1777–1791.

34.   MH Abraham, HS Chadha, GS Whiting, RC Mitchell. Hydrogen bonding. 32. An analysis of water–octanol and water–alkane partitioning and the $\Delta \log P$ parameter of Seiler. J Pharm Sci 83:1085–1100, 1994.

35.   MH Abraham, JC McGowan. The use of characteristic volumes to measure cavity terms in reversed phase liquid chromatography. Chromatographia 23:243–246, 1987.

36.   MH Abraham, R Kumarsingh, E Cometto-Muniz, WS Cain, M Roses, E Bosch, ML Diaz. The determination of solvation descriptors for terpenes, and the prediction of nasal pungency thresholds. J Chem Soc Perkin Trans 2 1998, 2405–2411.

37.   N El Tayar, R-S Tsai, B Testa, P-A Carrupt, A Leo. Partitioning of solutes in different solvent systems: The contribution of hydrogen-bonding capacity and polarity. J Pharm Sci 80:590–598, 1991.

38.   MH Abraham, F Martins, RC Mitchell, CJ Salter. Hydrogen bonding. 46. Characterization of the ethylene glycol–heptane partitioning system; hydrogen bond acidity and basicity of peptides. J Pharm Sci 88:241–247, 1999.

39.   HS Chadha, MH Abraham, RC Mitchell. Correlation and prediction of blood–brain distribution. J Mol Graphics 11:281–282, 1993.

40.   MH Abraham, HS Chadha, RC Mitchell. Hydrogen-bonding. 36. Determination of blood–brain distribution using octanol–water partition coefficients. Drug Design Discuss 13:123–131, 1995.

41.   MH Abraham, PK Weathersby. Hydrogen bonding. 30. The solubility of gases and vapours in biological liquids and tissues. J Pharm Sci 83:1450–1456, 1994.

42.   DJ Begley. The blood–brain barrier: Principles for targeting peptides and drugs to the central nervous system. J Pharm Pharmacol 48:136–146, 1996.

43.   F Lombardo, JF Blake, WJ Curatoio. Computation of brain–blood partitioning of organic solutes via free energy calculations. J Med Chem 39:4750–4755, 1996.

44.   U Norinder, P Sjoberg, T Osterberg. Theoretical calculation and prediction of blood–brain partitioning of organic compounds using MolSurf parametrization and PLS statistics. J Pharm Sci 87:952–959, 1998.

45.   JA Platts, MH Abraham, D Butino, A Hersey. J Chem Inf Comp Sci 39:835–849, 1999.

3

The Development of In Vitro Models for the Blood–Brain and Blood–CSF Barriers

Christiane Engelbertz, Dorothea Korte, Thorsten Nitz, Helmut Franke,* Matthias Haselbach, Joachim Wegener,** and Hans-Joachim Galla
University of Muenster, Muenster, Germany

I. INTRODUCTION

High amounts of new compounds of pharmaceutical interest are nowadays available by means of combinatorial chemistry. To develop their effectiveness in the central nervous system the substances have to cross barriers between blood and brain. Thus, besides binding to a specific receptor as the basis for their involvement in neuronal processes, permeation through the blood–brain barriers is a key event. Since thousands of new therapeutic compounds will have to be tested in the near future, alternatives to in vivo test systems must be developed. This is not only because of increasing ethical reasons but also simply because the handling of huge numbers of laboratory animals is extremely difficult and costly. Thus in vitro models that closely mimic the in vivo system, at least with respect to barrier properties, are in high demand.

* *Current affiliation*: Hoffman LaRoche, Basel, Switzerland.
** *Current affiliation*: Rensselaer Polytechnique Institute, Troy, New York.

Two barrier systems are of principal interest with respect to the passage of compounds into the brain. The blood–brain barrier in its original meaning is formed by a complex system of endothelial cells, astroglia, and pericytes, as well as the basal lamina interconnecting the cellular systems. The structural basis of this barrier consists of endothelial cells with tight junctions as special features that seal the intercellular cleft (1). Astrocytes, pericytes, and the extracellular matrix (ECM) components are believed to control the integrity of this barrier (2).

The second system that prevents the free passage of substrates between blood and brain is the barrier between the blood and the cerebrospinal fluid (CSF) built up by the epithelium of the choroid plexus. This epithelial barrier, again sealed by tight junctions, becomes indispensable, since the endothelium of choroid plexus capillaries is leaky and highly permeable to hydrophilic substrates (3).

In this chapter, in a state-of-the-art summary, we report progress in the establishment of in vitro models for both barriers. Figure 1 shows the in vivo situation for both barriers and the setup for filter experiments. We shall focus

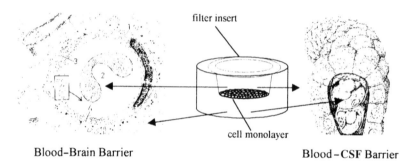

Blood–Brain Barrier   Blood – CSF Barrier

Fig. 1 Anatomical properties of the blood–brain barriers in vivo and the in vitro cell culture model. The blood–brain barrier is built up by the capillary endothelial cells sealed by tight junctions. Astrocyte endfeet (1) cover the endothelial surface, but extracellular matrix material (4) fills the intercellular cleft. Pericytes (3) are embedded in this basal membrane. An erythrocyte (2) is given for comparison. The vessel endothelium of the blood–CSF barrier is leaky, and the barrier function is taken by epithelial cells again sealed by tight junctions. In the cell culture model the lower filter compartment corresponds to the basolateral and the upper compartment to the apical side of the cells. It is important to note that the endothelial apical (luminal) side is the blood, whereas the apical epithelial side is the CSF side and the basolateral epithelial membrane is oriented to the blood. (Modified from Kristic, Die Gewebe des Menschen und der Säugetiere. Berlin: Springer-Verlag, 1982.)

on the expression of the barrier properties by withdrawal of serum from the culture medium. Hormonal effects and the influence of extracellular matrix components will be discussed. Since electrical resistance measurements are widely used to test the tightness of the barrier, we will scrutinize the different techniques now available.

II. PORCINE BRAIN CAPILLARY ENDOTHELIAL CELLS AS AN IN VITRO MODEL FOR THE BLOOD–BRAIN BARRIER

A. Preparation of Endothelial Cells and Culture Conditions

Porcine brain capillary endothelial cells (PBCEC) were isolated according to Bowman et al. (4) as modified by Tewes et al. (5). In brief, the meninges of freshly slaughtered pigs were removed to detach the larger blood vessels. Afterwards the cerebra were homogenized mechanically followed by enzymatic digestion in 1% (w/v) dispase II from *Bacillus polymyxa* dissolved in preparation medium (Earle Medium 199 supplemented with 0.7 mM L-glutamine, 100 µg/ml gentamicin, 100 U/ml penicillin, and 100 µg/ml streptomycin) for 2 hours at 37°C. Myelin and cell debris were separated by dextran density centrifugation at 4°C, 6800 g, for 10 minutes. The capillaries obtained were triturated by means of a glass pipet and digested with 0.1% (w/v) collagenase/dispase II in plating medium (preparation medium plus 10% v/v ox serum) at 37°C to remove the basement membrane. After 45 minutes the PBCEC were released. To purify the PBCEC from erythrocytes and any remaining myelin, a discontinuous Percoll density gradient was used (4°C, 1300 g, 10 min). The cell clusters obtained from one brain were sown on 500 cm^2 culture surface coated with collagen G or ECM protein. One day after sowing, cells were washed twice with phosphate-buffered saline containing Ca^{2+} and Mg^{2+} (0.8 mM each) and supplied with fresh growth medium (plating medium without gentamicin).

After 3 days PBCEC were passaged to Transwell® cell culture inserts (12 mm diameter, 0.4 µm pore size) in plating medium.

First the 1.13 cm^2 Transwell filter inserts were coated by adding 100 µl of different extracellular matrix protein solutions, which were diluted to 50 µg/ml protein shortly before use. The filter inserts were placed for 3 hours inside a sterile bench for drying. PBCEC were sown at 1.5×10^5 cells/cm^2 on the filter inserts. After 36 hours the medium was removed and replaced by culture medium: Dulbecco's modified Earle medium/Ham's F-12 with 6.5 mM L-glutamine, 100 µg/ml gentamicin, 100 U/ml penicillin, 100 µg/ml streptomycin, 550 nM hydrocortisone, 865 nM insulin, 30 nM sodium selenite, and 2 nM EGF (epidermal growth factor). In some experiments the culture medium was modified to investigate the influence of single supplements.

B. Cultures in Serum-Free Medium and the Effect of Hydrocortisone Introduction

Cell cultures have special demands for different factors that are normally supplied by the addition of sera (fetal calf serum, ox serum, etc.). Depending on the cells in culture, serum-containing media must be optimized because the serum composition is known with respect to the basic components only; precise knowledge with respect to growth factors like EGF or PDGF (platelet-derived growth factor), or hormones and metabolites is lacking. Since all these substances may interfere with cell proliferation and differentiation, it becomes difficult to study cellular processes in the presence of serum, especially if other cells like astrocytes or pericytes are involved (e.g., in the case of the differentiation process of cerebral capillary endothelial cells). Thus it is reasonable to make efforts to grow endothelial cells in serum-free medium, or at least to store them there. We recently reported that serum withdrawal drastically reinforces the blood–brain barrier properties of cultured endothelial cells (6). The extent of the expression of barrier properties depends on the addition of hydrocortisone, but also on plating density and ECM components.

Figure 2 shows the effect of serum withdrawal on electrical transendothelial resistance (TER). In the presence of serum, cells develop resistances not exceeding 200 $\Omega \cdot cm^2$. Exchange of this growth medium against chemically defined medium DMEM/Ham's F12 containing 10 ng/ml EGF, 30 nM selenite, 865 nM insulin, and 550 nM hydrocortisone increased the resistance up to 1000 $\Omega \cdot cm^2$. We tested all supplements, but only hydrocortisone was found to stimulate barrier properties in its physiological range (Fig. 3). Neither EGF nor selenite addition resulted in an increase in the TER; insulin was effective only at unphysiologically high concentrations. The effect depends drastically on the plating density (Fig. 4). We found an optimum around 1.5 × 10^5 cells cm^2. Since proliferation is strongly inhibited in the absence of serum, it is obvious that at low plating densities, cells do not reach confluency under serum-free conditions. Surprisingly, also, high cell densities resulted in low resistance values. We assume that because of the excess of cells, damaged cells that are not able to form tight junctions adhere to the filter surface. At best, the surface coverage may be so high that cells are still able to proliferate slightly. This seems to be just enough to seal the tight junctions.

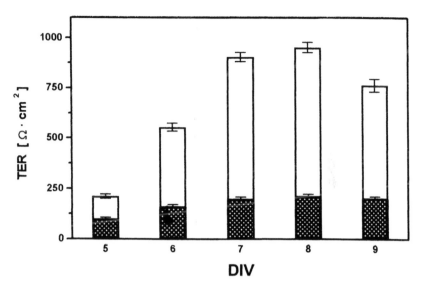

Fig. 2 The effect of serum withdrawal on transendothelial electrical resistances using porcine brain capillary endothelial cells (PBCEC) at different days in vitro (DIV). In the presence of serum (darkened part of each column), cells reach only low electrical resistances. Serum withdrawal allows the development of barrier properties with resistances of 1000 $\Omega \cdot$ cm^2 and above. (From Ref. 6.) Culture conditions: DMEM/Ham's F12 containing 2.5 mM L-glutamine, antibiotics, and EGF (10 ng/ml), selenite (30 nM), insulin (865 nM), and hydrocortisone (550 nM) with or without ox serum.

The effect of serum and hydrocortisone on electrical resistance and on sucrose permeability is summarized in Fig. 5. If the resistance in medium without serum and without hydrocortisone is set as the reference value (3 in Fig. 5), the addition of 550 nM hydrocortisone increases the TER by 150% (4 in Fig. 5). Serum addition in the presence of hydrocortisone yields a 25% decrease, whereas addition of serum in the absence of hydrocortisone decreases the TER by 50%. Sucrose permeability decreased correspondingly from 7×10^{-6} cm/s in the presence of serum and absence of cortisone, to 4×10^{-6} cm/s after cortisone addition down to 0.5×10^{-6} cm/s in serum-free medium containing 550 nM hydrocortisone (6).

C. Involvement of the Glucocorticoid Receptor

It is known that glucocorticoids reinforce barrier properties of cultured cell monolayers. Dexamethasone, a synthetic glucocorticoid, leads to a significant

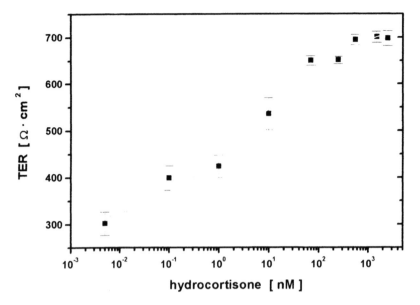

Fig. 3 The effect of hydrocortisone on the transendothelial electrical resistance of porcine brain capillary endothelial cells in the absence of serum. Cells reach strong barrier properties in the physiological hydrocortisone range around 500 nM. (From Ref. 6.)

increase in the TER of some cells like 31EG4 and rat mammary epithelial tumor cells. This effect correlates in a dose-dependent way with glucocorticoid receptor occupancy (7). Dexamethasone also results in a drastic increase in the TER of PBCEC, which even exceeds the effect of hydrocortisone by about 10% (data not shown).

We investigated whether hydrocortisone acts on PBCEC by the known pathway. In short, the water-insoluble steroid hormones are thought to cross the cellular plasma membrane by simple diffusion, bind tightly to a cytoplasmic protein, and cause this receptor protein to undergo a conformational change. In the form of a homodimer, two hormone–receptor complexes can bind to a specific DNA sequence (hormone response element) and thus regulate transcription or, more rarely, repress transcription of specific genes (8).

The antiprogestin-antiglucocorticoid RU38486 (mifepriston) was used to antagonize the effect of hydrocortisone. This synthetic substance presumably does not act by competing with the hormone for binding to the receptor

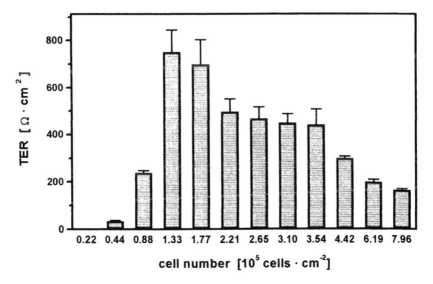

Fig. 4 Influence of plating density on the TER of PBCEC monolayers, which were cultivated in serum-free DMEM/Ham's F12 with supplements (see, e.g., Fig. 2). Plating medium was exchanged for cultivation medium one day after subcultivation. Cell number was determined by means of a Bürker hemocytometer. Experiments were performed after 8 days in vitro; the highest TER values were measured for all the cell densities. Data are given as mean $\pm$ SD ($n = 6$).

but by inducing or stabilizing a conformation of the ligand–receptor complex, which is different from the one induced by the agonist hydrocortisone or dexamethasone. As shown in Fig. 6 RU38486 in a 10-fold excess almost completely compensates the hydrocortisone-induced increase in the TER of PBCEC monolayers, which supports the assumption that hydrocortisone stimulates PBCEC via its specific cytosolic receptor.

D.   The Effect of Extracellular Matrix Proteins as Coatings

In vivo the capillary endothelial cells of the brain are surrounded by a specialized form of the extracellular matrix, the basement membrane (BM). The functions of BMs are diverse. They are separating layers that stabilize the histological pattern, act as a filter (e.g., the glomerula of the kidney), and mediate adhesion between cells and connective tissue matrix (9). Morphogenetic func-

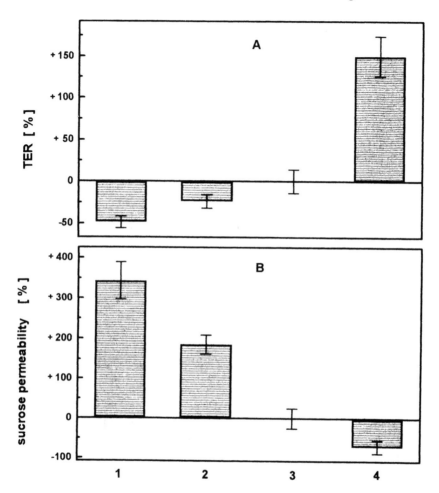

Fig. 5 Influence of hydrocortisone and ox serum on the barrier properties of PBCEC monolayers. Experiments were performed after 7 days in vitro. Cultivation media were as follows: 1, with 10% (v/v) ox serum, without hydrocortisone; 2, with 10% (v/v) ox serum, with 550 nM hydrocortisone; 3, without serum, without hydrocortisone; and 4, without serum, 550 nM hydrocortisone. Cell monolayers in the absence of serum and without hydrocortisone were taken as reference and set to zero (column 3 in A and B). (A) Analysis of the TER. Relative data are given as mean ± SD ($n = 36$). (B) Sucrose permeability. Relative data are given as mean ± SD ($n = 8$). (From Ref. 6.)

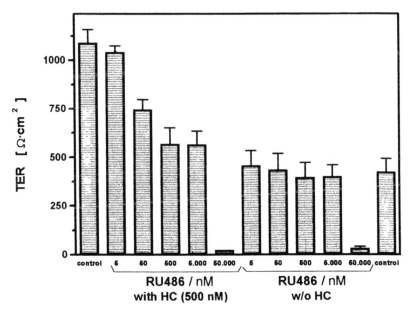

Fig. 6 Inhibition of the effect of hydrocortisone on the TER of PBCEC monolayers by RU38486 (mifepriston). Experiments were performed after 8 days in vitro. Cultivation medium: DMEM/Ham's F12 with 2.5 mM L-glutamine and antibiotics. Data are given as mean ± SD (n = 5).

tions are known during development. Thus the BM influences processes such as cell differentiation, proliferation, migration, adhesion, axon growth, and polarization. The BM consists of collagen type IV, laminin, entactin/nidogen, and heparan sulfate proteoglycan; some basement membranes contain further proteins like fibronectin or tenascin. The most abundant ECM protein is collagen. Type IV collagen (CIV) is known to build up a network in which the C-terminal NC 1 domain and the N-terminal 7S domain of the CIV monomers interact with each other during self-assembly. The 7S domains form tetramers by lateral alignment over 30 nm distance, alternating in a parallel and anti-parallel fashion (10). Hexamers are built up via the NC 1 domains (11). A cell-binding domain is present 100 nm away from the N-terminus. This area contains the recognition site for the two integrin receptors $\alpha_1\beta_1$ and $\alpha_2\beta_1$. Integrins serve as transmembrane bridges between the ECM and actin-containing filaments of the cytoskeleton. The connection between integrins and the actin cytoskeleton occurs in specialized structures known as focal adhesions. The integrin-mediated anchorage is a key regulator of apoptosis (12) but also plays an important role in signal transduction processes (e.g., by activation of tyrosine and serine/threonine kinases).

Another common BM protein is laminin (LM). LM has a cruciform structure and distinct cell-binding properties. During embryogenesis, LM appears as the first ECM protein. It also plays a role in cell polarization (10). LM has the ability to polymerize in vitro through interactions between the terminal globular domains of the molecule. LM can bind to heparan sulfate proteoglycan, and to CIV. The CIV binding is effectively mediated by nidogen. The binding to cell surfaces is mediated via integrins like $\alpha_6\beta_1$ or $\alpha_1\beta_1$, via non-integrin-binding proteins, and via carbohydrate-binding moieties such as lectins (13). Another important ECM protein is fibronectin (FN). FN contains binding sites for ECM proteins such as collagen and thrombospondin, glycosaminoglycans such as heparin and chondroitin sulfate, circulating blood proteins such as fibrin, and cell surface receptors such as the $\alpha_5\beta_1$ integrin (14). FN is a ubiquitous glycoprotein that exists in a soluble form in body fluids and in an insoluble form in the extracellular matrix. It influences diverse processes including inflammation, wound repair, malignant metastasis, microorganism attachment, and thrombosis (15).

We have investigated the possible involvement of extracellular matrix proteins in PBCEC tight junction formation (16). In particular, the influence of basement membrane proteins on the transcellular electrical resistance of confluent PBCEC monolayers—as a measure of ionic permeability of the intercellular tight junctions—should be clarified. Rat tail collagen, shown previously to be a suitable substratum for PBCEC cultivation on filter inserts (5), consists mainly of type I collagen, which does not occur in basement membranes, hence could be used as a reference base.

Comparison of TER values obtained from different PBCEC primary cultures demonstrated that the cell layer's electrical resistance on a rat tail collagen substratum varied from approximately 200 $\Omega \cdot cm^2$ to 1100 $\Omega \cdot cm^2$ (see Fig. 7). This variability of TER is correlated to a given cell preparation. From the large number of experiments, two groups of cell monolayers became apparent, one containing five preparations (26 filters) exhibiting a relatively low TER (around 350 $\Omega \cdot cm^2$), and a second group of four preparations consisting of 19 filters peaking around 1000 $\Omega \cdot cm^2$.

For further experiments, PBCECs were cultured on filter inserts coated with type IV collagen, laminin, fibronectin, and 1:1 (mass ratio) mixtures of these proteins. TERs of cells on basement membrane protein substrata were compared to TERs on rat tail collagen. As mentioned above, PBCEC monolayers on rat tail collagen displayed resistances of approximately 350 or 1000

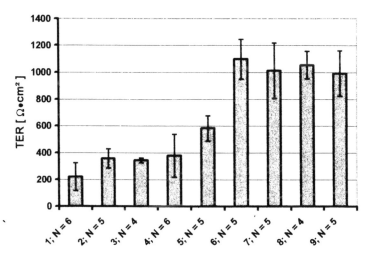

Fig. 7 TER and PBCEC monolayers on filter inserts coated with rat tail collagen derived from nine individual preparations; N is the number of filter inserts per preparation. Values are given as mean ± SD (bars). (From Ref. 16.)

$\Omega \cdot cm^2$, depending on the preparation. Consequently, we decided to quantify basement membrane protein influence on TER in two separate evaluations: one for PBCEC cultures exhibiting low TER on rat tail collagen and another one for PBCEC whose TERs on rat tail collagen were already high.

Type IV collagen, fibronectin and laminin as well as 1:1 mixtures of them elevated TER of PBCEC compared to the rat tail collagen reference base (Fig. 8). This effect was, however, correlated to the TER level on rat tail collagen. If the resistance on rat tail collagen was low (i.e., around 350 $\Omega \cdot cm^2$), a 2.3-fold (laminin) to 2.9-fold (fibronectin/laminin) increase was observed (Fig. 8A). In contrast, TER of PBCEC reaching 1000 $\Omega \cdot cm^2$ on rat tail collagen could be only slightly elevated by basement membrane proteins (Fig. 8B). The increase in TER observed on fibronectin, laminin, and type IV collagen/laminin was not significant. A small but significant elevation of electrical resistance could be detected on type IV collagen, type IV collagen/fibronectin (1.1-fold), and fibronectin/laminin (1.2-fold).

These observations show that transcellular electrical resistances of cerebral capillary endothelial cells may be influenced by basement membrane proteins. In particular, type IV collagen, fibronectin, and laminin, as well as mixtures of these proteins, significantly elevated TER of porcine brain capillary endothelial cells in vitro. The results suggest that type IV collagen, fibronectin,

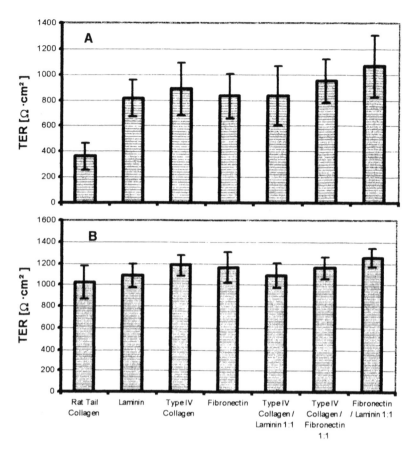

Fig. 8 Effect of basement membrane proteins on TER of PBCEC monolayers. Data obtained from three PBCEC preparations are displayed as mean ± SD (bars). (A) PBCEC monolayers with low resistance values on rat tail collagen. Data were taken from preparations 2, 3, and 4 in Fig. 7. Number of filter inserts per substratum, N: 16 for rat tail collagen and type IV collagen, 17 for fibronectin and fibronectin/laminin, and 18 for laminin, type IV collagen/fibronectin, and type IV collagen/laminin. (B) PBCEC monolayers with high resistance values on rat tail collagen. Data were taken from preparations 7, 8, and 9 in Fig. 2. Number of filter inserts per substratum, N: 14 for rat tail collagen, 17 for fibronectin, 21 for type IV collagen, and 18 for all other substrata. (From Ref. 16.)

and laminin are also involved in tight junction formation between endothelial cells. One possible explanation is that these extracellular proteins provide an informational cue that is transmitted to the inside of the cell either mechanically by the cytoskeleton or chemically via signal transduction pathways directing tight junction proteins to the apical–basolateral boundary. On the other hand, proteins like type IV collagen and laminin may also bind to their cellular receptors and activate intracellular signal transduction pathways leading to the expression of tight junction–associated proteins.

If, however, cultured PBCECs form layers with an already high TER of 1000 $\Omega \cdot$ cm^2 even on rat tail collagen, they are obviously at a more mature stage of differentiation, which cannot be influenced by extracellular matrix proteins to the same extent. These observations on differential TER elevation are particularly important for primary cultures of cerebral endothelial cells, where significant variations of cell quality can occur. It must be pointed out that the use of basement membrane proteins as a growth substrate decreases the differences between individual endothelial cell preparations, an effect that is important for the use of primary cell cultures as in vitro models.

III. THE BLOOD–CSF BARRIER IN VITRO

A. Cultivation of Choroid Plexus Epithelial Cells

Epithelial cells from porcine choroid plexus were obtained by a modified preparation basically described by Crook et al. (17) with minor modifications according to Gath et al. (18). Briefly, the choroid plexus tissue was removed from porcine brains and incubated with 0.25% trypsin solution for 2.5 hours at 4°C. During this time the trypsin, inactive at 4°C, could penetrate the tissue. Thereafter the solution was warmed up to 37°C for 30 minutes. After the trypsin digestion had been stopped by addition of newborn bovine serum, the enzymatically released cells were pelleted by centrifugation and seeded on laminin-coated permeable filter inserts. The laminin coating ensured the formation of confluent monolayers because of an improved adherence and proliferation of the epithelial cells on laminin.

The epithelial cells were cultivated in DME/Ham's F12 medium with several supplements (10% fetal bovine serum, 4 mM L-glutamine, 5 µg/ml insulin, 100 µg/ml penicillin, and 100 mg/ml streptomycin) and, most important, 20 µM cytosine arabinoside. Cytosine arabinoside is a nucleoside containing arabinose as a sugar. This molecule, which is also used in cancer therapy, is not a substrate for the transport system for desoxyribonucleosides and ribonucleosides of the choroid plexus epithelial cells (19). Other cells (e.g., fibroblast-like cells, which are also released by the trypsin digestion from the choroid plexus tissue) do not distinguish between nucleosides with ribose or arabinose as sugar residues. Thus the incorporation of cytosine arabinoside into the RNA of contaminating cells transcriptionally prevents their proliferation and guarantees the cultivation of pure choroid plexus epithelial cell cultures. Since the treatment with cytosine arabinoside did not affect the growth and morphology of the epithelial cells (18), we were able to establish an in vitro model of the blood–CSF barrier without contaminating cells on permeable filter inserts. In this model system the apical compartment of the filter represents the ventricular side in vivo, and the basolateral side with the laminin coating represents the side facing the basal membrane and the fenestrated choroid plexus capillaries (Fig. 1).

B. Application of Serum-Free Medium

1. Morphological Changes

Excellent cultivation conditions for epithelial cells were achieved in the presence of FBS, but some cells showed a restricted polarity that was easily recognizable, for example, by the truncated microvilli. In vivo the ventricular surface of the epithelial cells is covered densely and homogeneously with microvilli (brush border). The microvilli have several functions: they enlarge the surface area for absorption, transport, and diffusion processes; and they help distributing transported proteins, nucleosides, and ions by displaying motility. Cells cultivated in the presence of serum do have microvilli, but they are heterogeneously distributed, and most of them display a truncated shape (Fig. 9a and c). Morphological changes take place after serum withdrawal. The apical surface of the epithelial cells is characterized by a homogeneous distribution of fully developed microvilli (Fig. 9b and d). Thus the cell polarity with respect to microvilli is improved in the absence of serum.

2. Improved Enzyme Polarity

Another hint for the improved polarity of epithelial cells in the absence of serum is the increased expression of Na^+,K^+-ATPase, an enzyme that is indispensable for the active transport processes and the liquor production (see below). It is located at the apical membrane. Besides the epithelial cells of the choroid plexus, the apical localization of Na^+,K^+-ATPase could be shown only for the corneal epithelial cells (20). In serum-free medium the expression of Na^+,K^+-ATPase is significantly increased (Fig. 10). This could be shown by immunocytochemical studies as well as by quantification

Fig. 9 Scanning electron microscopic picture of choroid plexus epithelial cells in the presence of fetal bovine serum (left panels) and in the absence of serum (right panels). Note the increased number and the fully developed microvilli after serum withdrawal. (From Ref. 21.)

of the Na^+,K^+-ATPase activity (21). The immunocytochemistry reveals a homogeneous staining of cells cultivated in serum-free medium for 4 days. In contrast, cells cultivated in serum-containing medium only show an irregular expression; some cells are hardly stained. This observation was confirmed by means of quantitative analysis of the Na^+,K^+-ATPase activity (Stolz and Jacobson (22)). The activity of the enzyme with respect to total protein is increased about two-fold after serum withdrawal for 4 days (Fig. 10). The activity reaches the same level that was obtained in freshly prepared epithelial cells. Again, the removal of serum affects the cell polarity. This differentiation process of the choroid plexus cells in serum-free medium may be similar to the maturation process of the developing choroid plexus tissue in vivo. As Parmelee and Johanson (23) have shown, the Na^+,K^+-ATPase activity is significantly higher in the adult choroid plexus

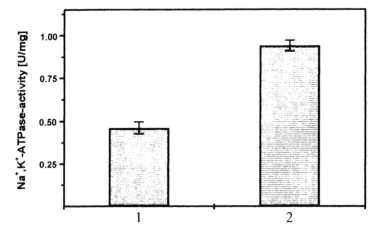

Fig. 10 Expression of Na$^+$,K$^-$-ATPase in choroid plexus epithelial cells in the presence (1) and absence (2) of serum. (From Ref. 21.)

than in the infant tissue. One possible explanation is that the fetal serum contains substances that prevent the complete differentiation of the epithelial cells in vivo and in vitro. After withdrawal of serum, the differentiation processes can be completed.

3. Improved Barrier Properties

For the establishment of a blood–CFS barrier model, confluent monolayers should exhibit physiological functions. First of all, a low permeability for high molecular substances needs to be achieved. This means also improved cell–cell interactions (tight junctions), which can be observed by the measurement of the transepithelial electrical resistance (TER). Confluent choroid plexus epithelial monolayers cultivated in the presence of serum display a TER of 150 $\Omega \cdot cm^2$ (18). This is in good agreement with reported resistances of 100 $\Omega \cdot cm^2$ when rat choroid plexus epithelial cells were used (24). Remanthan et al. (25) cultivated choroid plexus epithelia of rabbit and measured transepithelial resistances of 50–80 $\Omega \cdot cm^2$. Changes in TER values of the porcine epithelial cells could be observed in serum-free medium. After withdrawal of serum for 4 days, a dramatic increase of the TER from 150 $\Omega \cdot cm^2$ to 1250 $\Omega \cdot cm^2$ is reached (21). After 8 days without serum, an even larger TER of 1700 $\Omega \cdot cm^2$ is reached (Fig. 11). TER values higher than 1500 $\Omega \cdot cm^2$ do not correlate to published data derived from in vivo measurements of choroid plexus tissue.

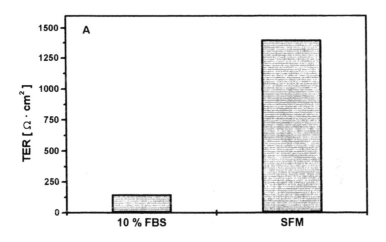

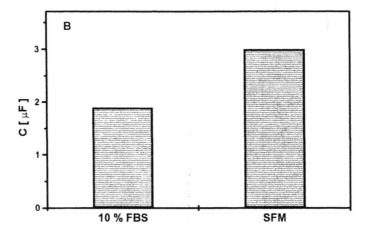

Fig. 11 Transepithelial resistance (A) and capacitance (B) of the epithelial mono-layer determined by impedance spectroscopy in the presence and absence of serum: FBS, fetal bovine serum; SFM, serum-free medium. (From Ref. 21.)

Zeuthen and Wright (26) reported that the TER of freshly isolated choroid plexus tissue of the bullfrog exhibits a low value of about 25 $\Omega \cdot cm^2$. However, these extremely low values are not compatible with the high complexity of the tight junction strands. Since physiological activity of the porcine epithelial cells (see below) was observed only for cells with dramatically increased TER values, we assume that TER values of 1700 $\Omega \cdot cm^2$ reflect the physiological in vivo situation of the mammalian plexus epithelium rather than the low TER values determined in the amphibian choroid plexus tissue.

By means of impedance spectroscopy, we are able to determine not only the resistance of a given monolayer but also the capacitance. The capacitance is mainly characterized by the plasma membrane surface area. In serum-free medium, the cell layer's capacitance is also increased. This is due to the increased number and extension of the microvilli, a change that causes the apical membrane to increase in area. Changes in basolateral membrane area or changes in protein content and distribution are not excluded by our experiments and might also affect the capacitance of a cellular membrane. But these possible changes are rather unlikely to be mainly responsible for the observed increase in capacitance (Fig. 11). In summary, the absence of serum improves not only the polarity of the plexus epithelial cells but also the cell–cell interactions. This can be clearly seen by the rise in TER values corresponding to the development of dense tight junction strands.

C. Secretion of Cerebrospinal Fluid

The cerebrospinal fluid (CSF) helps to maintain a stable and specialized fluid environment for neurons. The main source of CSF are the choroid plexuses, generating 75% or more of the total fluid formed (3). As an active secretion, CSF is not simply the result of a filtration of fluid across membranes; rather, it is a process driven by the active transport of ions from the blood into the ventricular system. Since the main constituents of the CSF are Na^+, Cl^-, and HCO_3^-, knowledge of how these ions are transported is essential for an understanding of the key elements of CSF production.

The driving force for CSF secretion is the Na^+,K^+-ATPase in the apical cell membrane, which decreases the intracellular concentration of Na^+ and increases it extracellularly. The resulting inwardly directed Na^+ gradient promotes basolateral Na^+/H^+ exchange and Na^+,Cl^- cotransport also at the blood-facing basolateral membrane. Furthermore Cl^- uptake is driven by exchange with intracellular HCO_3^- generated from CO_2 hydration, which is catalyzed by carbonic anhydrase (Fig. 12).

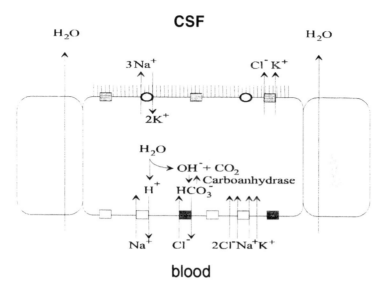

Fig. 12 Biochemical mechanism of the fluid secretion of plexus epithelial cells; details in the text.

At the CSF side of the cell, Na^+ is actively pumped into the ventricles. In addition, K^+ and Cl^- leave the cells via channels in the apical membranes, as does HCO_3^- at the basolateral side. Through the net flux of Na^+ and accompanying anions across the entire choroidal membranes, water, which comprises 99% of CSF, is forced by the osmotic gradient to move from the blood into the ventricles.

Under serum-free conditions, cultured choroid plexus epithelial cells transport fluid from the basolateral to the apical cell side. Figure 13 demonstrates the increased level of the apical medium. Fluid secretion requires as the driving force both high transmembrane resistance and a polar distribution of the Na^+,K^+-ATPase at the apical cell surface. Under these conditions, cells are able to build up ion gradients across the cell monolayer. The concentration of electrolytes such as Na^+ and Cl^- in the apical compartment is expected to be higher than it is in the basolateral compartment as a consequence of the Na^+,K^+-ATPase activity at the apical membrane side. This leads to a significant water flow across the cell monolayer into the apical compartment. Fluid secretion can be inhibited by the addition of ouabain, proving the correlation between Na^+,K^+-ATPase activity and fluid secretion in cultured cells. Such inhibition of fluid secretion was mainly observed after application of ouabain

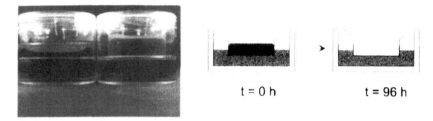

Fig. 13 Fluid secretion of cultured plexus epithelial cells under serum-free conditions: left filter, without cells; right filter, confluent cell layer 4 days after serum withdrawal. The filter medium is stained with phenol red, a substrate of the organic anion transporter. Note that parallel to the increased fluid level at the apical compartment the dye has been transported from the apically oriented filter compartment to the basolateral compartment. (From Ref. 21.)

from the apical membrane side corresponding to the known apical localization of Na^+,K^+-ATPase in choroid plexus epithelial cells (27).

The fluid secretion in the cell culture model is a saturable process correlated with the formation of a proton gradient across the cell monolayer. The pH in the basolateral chamber significantly decreases as a function of time corresponding to an increase of the pH in the apical chamber. Since the Na^+,H^+ exchanger transports protons out of the cells into the basolateral chamber, the increasing proton concentration could inhibit the exchange. This would consequently inhibit the Na^+,K^+-ATPase at the apical membrane side as both transport systems are functionally coupled. Inhibition of Na^+,K^+-ATPase would then lead to a decreased fluid secretion which has been observed after ouabain inhibition. This corresponds to the finding that higher concentrations of $NaHCO_3$ resulting in a higher buffer capacity of the incubation medium significantly lead to an increased fluid secretion (27).

D. Active Transport Properties

Choroid plexus cells display a central role in the regulation of brain homeostasis (28). Active transport systems located in the apical and basolateral membrane side of the epithelial cells regulate both the secretion and the composition of the cerebrospinal fluid. Since substances in the CSF have free access to the brain tissue, the choroid plexus is mainly responsible for the micronutrient homeostasis in the brain. Vitamins like ascorbic acid and myoinositol are actively transported from blood into the brain, whereas substrates like riboflavin and penicillin G are cleared out of the brain into the blood.

In the cell culture model, active transport across the cell layer was first observed by the clearance from the apical to the basolateral chamber of phenol red, which was supplemented to the culture medium as pH indicator. This transport process against the concentration gradient requires the low permeabilities obtainable only in serum-free medium. Otherwise back-diffusion of the substrates would compensate for any concentration gradient. The transport for phenol red is almost linear until 30% of the initial amount in the apical chamber has been transported against the concentration gradient into the basolateral compartment (27).

The transport of penicillin G, riboflavin, and fluorescein from the brain into the blood by the organic anion exchanger was reported in 1997 (29). Investigations using the cell culture model showed that beside these substrates, phenol red is also a ligand of this transport system in the choroid plexus. Figure 14 shows the competitive inhibition of the phenol red transport by penicillin G.

It was shown earlier that the organic anion transporter requires an outwardly directed chloride gradient at the apical membrane side. This gradient

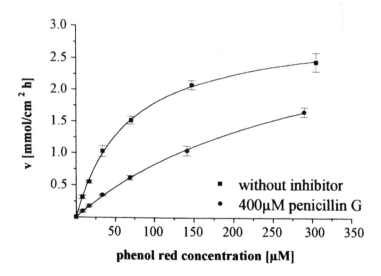

Fig. 14 Phenol red transport across plexus epithelial cell monolayers. The transport velocity is given as function of phenol red concentration in the basolateral compartment. The dye transport is inhibited by penicillin G. (From Ref. 27.)

Table 1 Kinetic Data for the Active Transport Systems Investigated

Substrate	Transport direction[a]	Michaelis constant, K_m (μM)	V_{max} (nmol h^{-1} cm^{-2})	Value of K_m determined by uptake measurements (μM)	Ref.
Ascorbic acid	b → a	67 ± 12	3.91 ± 0.29	44	10
Myoinositol	b → a	117 ± 9	1.65 ± 0.05	≈100	11
Penicillin G	a → b	107 ± 8	1.82 ± 0.05	58–70	17,16
Fluorescein	a → b	22 ± 1.5	1.92 ± 0.05	40	2
Phenol red	a → b	68 ± 2.9	3.01 ± 0.06	Not observed so far	
Phenol red + 400 μM pencillin G		326 ± 18	3.50 ± 0.12		
Riboflavin	a → b	78 ± 4	1.84 ± 0.05	78	14

[a] a, apical; b, basolateral.

is mainly built up by the Cl^-/HCO_3^- exchanger activity. The stilbene derivative SITS inhibits this exchange system and consequently also inhibits the transport of substances by the organic anion exchanger. In the cell culture system, SITC decreased the transport rate of organic anions up to 60%. Since the transport of riboflavin is inhibited by SITS to a lower extent than is the transport of fluorescein and phenol red, it can be assumed that the organic anion transporter might not be the only transport system involved in riboflavin transport across the choroid plexus cell monolayer (27).

Table 1 shows the kinetic data of the investigated transport systems of the cultured choroid plexus epithelial cells.

IV. SCRUTINY OF ELECTRICAL RESISTANCE MEASUREMENTS

One method of quantifying cellular barrier formation is to measure transepithelial or transendothelial resistance, which is mainly regarded as a direct measure of ion permeability. However, large discrepancies in resistance measurements have been reported, and their significance is presently under discussion. On the one hand, differences in the type of cell culture or species differences (e.g., heterogeneities in the cell monolayer of contaminating cells that do not form tight junctions) may be responsible for the low reproducibility and comparability of published results. On the other hand, it is now accepted that different techniques yield different results. This is clearly demonstrated in Fig. 15, which shows resistances of porcine brain capillary endothelial cells from the same culture in the absence and presence of a cyclic AMP analogue obtained by means of either the so-called chopstick electrodes STX-2 or the Endohm-12 or Endohm-24 (for filters with 12 or 24 mm diameter). Both setups were equipped with the Millicell ERS® volt-ohmmeter applying a rectangular current pulse of ± 20 μA with a frequency of 12.5 Hz. The STX-2 system uses two pairs of locally fixed current and voltage electrodes, one of each pair of sticks being in the lower filter compartment, the other in the upper. In the Endohm system, the current and the voltage electrodes are arranged in a concentric way with fixed positions and large enough areas to cover the whole filter and to allow a fixed gap adjustment. The results given in Fig. 15 clearly show the discrepancies. Values obtained with the chopstick setup were not very reproducible and exhibited deviations up to ± 50%, depending on the position of the electrodes and their distance from the filter, the lateral position, and the angle of submersion even when the same cell-covered filter was used.

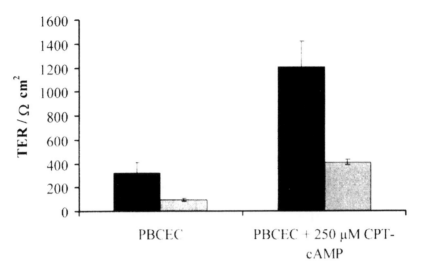

Fig. 15 Comparison of resistance measurements made by means of chopstick electrodes STX 2 (dark columns) and by means of the Endohm system (lighter columns). Porcine brain capillary endothelial cells were cultured in the presence of serum without (left-hand pair of columns) a membrane-permeating cAMP analogue and after addition of such a substance (right-hand pair).

Moreover, all values measured with the chopstick arrangement were significantly higher than those measured with the Endohm setup.

As shown in Fig. 15 in the absence of CPT-cAMP, typical values in serum-containing media were $324 \pm 88 \ \Omega \cdot cm^2$ for the chopstick and $96 \pm 14 \ \Omega \cdot cm^2$ for the Endohm setup. In the presence of CPT-cAMP we measured $1206 \pm 210 \ \Omega \cdot cm^2$ with the STX-2 electrodes and $412 \pm 20 \ \Omega \cdot cm^2$ with the Endohm equipment. We further observed that the Endohm 24 system focuses the measurement on the center part of the filter and thus yields high resistances even if the cell monolayer shows defects close to the edge of the filter. When the smaller diameter Endohm-12 chamber was used, reliable and reproducible results were obtained. This already shows that more sophisticated and more reliable methods are needed to quantify electrical resistances of barrier-forming cells.

Progress was obtained with a small-area device for resistance measurements in work reported by Erben et al. (30). To measure individual colonies of cell monolayers, the device for resistance measurement is attached to the microscope stage with two electrodes (one for the current, one for the voltage) placed on an upper filter holder carrying the cell culture insert, whereas the lower chamber with voltage and current electrodes is placed on the microscope objective. A narrow tube on top of the lower compartment approaches the filter and thus determines with its inner hole the area of measurement down to 1 mm diameter. This setup yields resistance values in good agreement with the Endohm-12 measurement. This method, however, was limited experimentally to resistances not exceeding 500 $\Omega \cdot cm^2$, which is not sufficient for expected values close to 2000 $\Omega \cdot cm^2$ from reported in vivo data.

The most advanced method available today is impedance spectroscopy (IS), a powerful technique to investigate electrical properties of surface materials including cells (31–35). IS yields information about both conductivity and dielectric constant (capacitance) of the interfacial region, which in our case is the cell monolayer (see Fig. 11). For impedance measurements, an ac sinusoidal potential U_0 is applied across the cell monolayer by means of two electrodes, and the current $I(f)$ is measured. The ratio $U_0/I(f)$ and the phase shift $\varphi(f)$ between potential and current is recorded as function of frequency between 1 Hz and 1 MHz, giving the impedance spectrum $|Z_f|$. To obtain the values of electrical parameters, the transfer function of the equivalent circuit is determined and fitted to the experimental data. We used an impedance analyzer SI1260 (Solartron Instruments, U.K.) with a built-in frequency generator operating in the frequency range between 10 μHz and 32 MHz. The amplitude of the applied ac potential was adjusted to 10 mV.

Two different ways to apply the ac potential have been used (see Fig. 16): (a) cells are grown directly on top of a thin gold surface that served as measuring electrode, and (b) a cell-covered filter is positioned above a glass slide covered with an evaporated gold film as working electrode. A circular platinized electrode in the compartment above the cell monolayer served as the counter electrode. The first setup is used for resistance and capacitance measurements only; the second setup is used if resistance measurements are correlated to permeability determinations. Typical impedance spectra $|Z_f|$ of gold electrodes with and without cells are given in Fig. 17. Barrier-forming plexus epithelial (Fig. 17A) cells are compared with aortic endothelial cells (Fig. 17B), developing weak barrier properties. This new technique can be used to measure transcellular electrical resistance from very low values ($<$ 10 $\Omega \cdot cm^2$) up to high values of several thousand ohm-square centimeters. Typical applications are shown in Fig. 18.

In addition to the electrical resistance measurements, this technique allows us to quantify the capacitance of the cell monolayer, which mainly reflects the passive electrical properties of the plasma membrane. Figure 11 showed the increase in resistance of choroid plexus epithelial cells induced

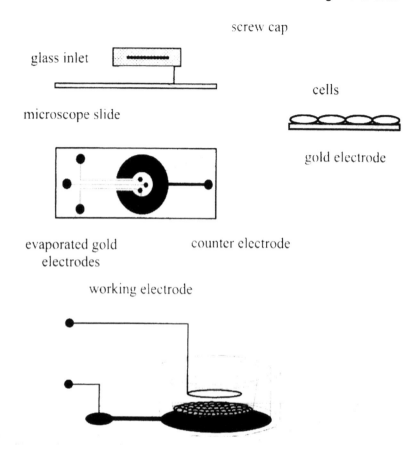

Fig. 16 Setup for making resistance and capacitance measurements by means of impedance spectroscopy. Cells are grown directly either on the fold electrodes or on filters. The second setup allows parallel determination of electrical resistance and substrate permeability.

by the exchange of serum-containing medium against serum-free medium. Parallel to the drastic increase in the electrical resistance, we observed an increase in capacitance as well. This increase in capacitance was assigned to an increased number and an extension of the epithelial microvilli, leading to an increased area of the apical membrane. This experiment again clearly shows that impedance spectroscopy is a powerful method yielding not only reproducible and reliable resistance values over a broad range from very low to very

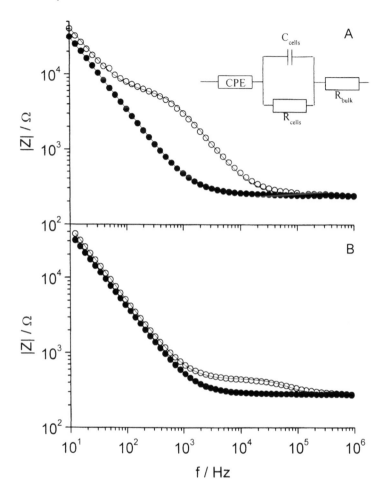

Fig. 17 Impedance spectra of (A) confluent plexus epithelial cells and (B) low resistance bovine aortic endothelial cells: open circles, cell monolayer on the gold electrode; solid circles, gold electrode after removal of the cells. The lines give the fit of
the transfer function of the equivalent circuit. Typical values are R_{bulk} = 250–300
$\Omega \cdot cm^2$. The resistance values are R_{cell} = 156 $\Omega \cdot cm^2$ for the plexus cells, here measured in the presence of medium, and R_{cell} = 5.1 $\Omega \cdot cm^2$ for the aortic cells. The
corresponding capacitances are C_{cell} = 1.7 $\mu F/cm^2$ for the plexus epithelial and C_{cell} =
0.6 $\mu F/cm^2$ for the aortic endothelial cells. The constant phase element (CPE) is an
empirical constant that corrects for nonidealities of the capacitance caused by the in-
homogeneities of the electrical interphase and the corresponding electrochemical
properties.

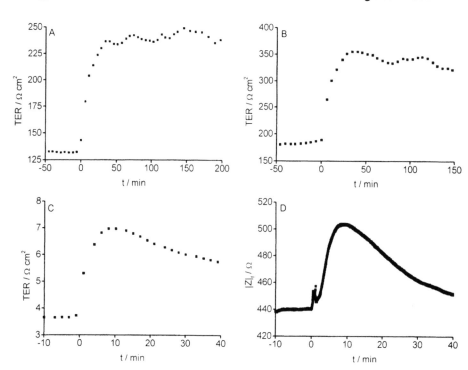

Fig. 18 Typical applications for impedance spectroscopy. (A) Time dependence of the electrical resistance of plexus epithelial cells in the presence of serum after addition of the CPT-cAMP analogue (100 μM) to increase the barrier properties of the cell monolayer. Complete spectra were taken and analyzed. (B) The same as (A) but the intercellular cAMP level was increased by addition of 2.5 μM forskolin. (C) Time-dependent change of the transcellular resistance of bovine aortic endothelial cells after stimulation with 1 μM isoproterenol, an unspecific β_1,β_2-receptor agonist. The expected increase is observable despite the low resistance values of these leaky endothelial cell monolayers. Note that only impedance spectroscopy is able to measure these low values accurately. (For further details, see Ref. 35.) (D) Time dependence of the impedance measured at a fixed frequency of $f = 10$ kHz after 100 μM CPT-cAMP stimulation of bovine aortic endothelial cells. If only relative changes are required, instead of an analysis of electrical resistance, impedance at a given frequency is an easy and fast method of following cellular responses.

high values but, in addition, information about morphological changes of the corresponding cell culture. To summarize: for pure resistance measurements the Endohm systems, small-area devices, and impedance spectroscopy yield identical values. Chopstick electrodes should be avoided owing to the practical shortcomings mentioned above. For more sophisticated applications, impedance spectroscopy is highly recommended.

V. CONCLUSIONS

Blood–brain barrier models now available make use of cerebral capillary endothelial or choroid plexus epithelial cells. Both cell types need serum in the growth medium to proliferate. Serum, however, inhibits the formation of tight cell–cell contacts. Withdrawal of serum favors cellular polarity and increases the barrier properties drastically. Electrical resistance is an easy measure of junctional tightness. Use of chopstick electrodes is not recommended, but the Endohm setup yields reliable results. A reported small-area device is applicable but not easy to handle. The seal between measuring capillary and filter is the critical parameter. A very sophisticated but highly reliable and reproducible new method is impedance spectroscopy, in which ac potentials are applied over a wide frequency range. At a single fixed frequency, ac potentials may be applied and analyzed if only relative changes after substrate application are expected.

REFERENCES

1. J Wegener, H-J Galla. The role of non-lamellar lipid structures in the formation of tight junctions. Chem Phys Lipids 81:229–255, 1996.
2. CT Beuckmann, H-J Galla. Tissue culture of brain endothelial cells—Induction of the blood–brain barrier properties by brain factors. In: W. Pardridge, ed. Introduction to the Blood–Brain Barrier. Cambridge: Cambridge University Press, 1998, pp 79–85.
3. CE Johanson. Ventricles and cerebrospinal fluid. In: P Michael, ed. Neuroscience in Medicine. Philadelphia: JB Lippincott, 1995, pp 171–196.
4. PD Bowman, SR Ennis, KE Rarey, AL Betz, GW Goldstein. Brain microvessel endothelial cells in culture: A model for study of blood–brain barrier permeability. Ann Neurol 14:396–402, 1983.
5. B Tewes, H Franke, S Hellwig, D Hoheisel, S Decker, D Griesche, T Tilling, J Wegener, H-J Galla. Preparation of endothelial cells in primary cultures obtained from the brains of 6-month-old pigs. In: AG de Boer, W Sutanto, eds.

Transport Across the Blood–Brain Barrier: In Vitro and In Vivo Techniques. Amsterdam: Harwood, Academic Publishers, 1997, pp 91–97.

6. D Hoheisel, T Nitz, H Franke, J Wegener, H-J Galla. Hydrocortisone reinforces the blood–brain barrier properties in a serum free cell culture system. Biochem Biophys Res Commun 244:312–316, 1998.

7. P Buse, PL Woo, DB Alexander, HH Cha, A Reza, ND Sirota, GL Firestone. Transforming growth factor-α abrogates glucocorticoid-stimulated tight junction formation and growth suppression in rat mammary epithelial tumor cells. J Biol Chem 270:6505–6514, 1995.

8. JW Funder. Glucocorticoid receptors and mineralocorticoid receptors: Biology and clinical relevance. Annu Rev Med 48:231–240, 1997.

9. HJ Merker. Morphology of the basement membrane. Microsc Res Tech 28:95–124, 1994.

10. M Weber. Basement membrane proteins. Kidney Int 41:620–628, 1992.

11. K Kühn. Basement membrane (type IV) collagen. Matrix Biol 14:439–445, 1994.

12. A Howe, AE Aplin, SK Alahari, RL Juliano. Integrin signaling and cell growth control. Curr Opinion Cell Biol 10:220–231, 1998.

13. L Luckenbill-Edds. Laminin and the mechanism of neuronal outgrowth. Brain Res Rev 23:1–27, 1997.

14. JR Potts, ID Campbell. Fibronectin structure and assembly. Curr Biol 6:648–655, 1994.

15. DJ Romberger. Fibronectin. Int J Biochem Cell Biol 29:939–943, 1997.

16. T Tilling, D Korte, D Hoheisel, H-J Galla. Basement membrane proteins influence brain capillary endothelial barrier function in vitro. J Neurochem 71:1151–1157, 1998.

17. RB Crook, H Kasagami, SB Prusinger. Culture and characterization of epithelial cell from bovine choroid plexus. J Neurochem 37:845–854, 1981.

18. U Gath, A Hakvoort, J Wegener, S Decker, H-J Galla. Porcine choroid plexus cells in culture: Expression of polarized phenotype, maintenance of barrier properties and apical secretion of CSF-components. Eur J Cell Biol 74:68–78, 1997.

19. R Spector. Pharmacokinetics and metabolism of cytosine arabinoside in the central nervous system. J Pharmacol Exp Ther 222:1–6, 1982.

20. C Simmons. The development of cellular polarity in transport epithelial. Semin Perinatol 16:78–89, 1992.

21. A Hakvoort, M Haselbach, J Wegener, D Hoheisel, H-J Galla. The polarity of choroid plexus epithelial cells in vitro is improved in serum-free medium. J Neurochem 71:1141–1150, 1998.

22. DB Stolz, BS Jacobson. Examination of extracellular membrane protein polarity of bovine aortic endothelial cells in vitro using cationic silica microbead membrane isolation procedure. J Cell Sci 103:39–51, 1992.

23. JT Parmelee, CE Johanson. Development of potassium transport capability by choroid plexus of infant rats. Am J Physiol 256, R786–R791, 1989.

24. BR Southwell, W Duan, D Alcorn, C Brack, SJ Richardson, J Köhrle, G

Schreiber. Thyroxine transport to the brain: Role of protein synthesis by the choroid plexus. Endocrinology 133:2116–2126, 1993.

25. VK Ramanthan, SJ Chung, KM Giacomini, CM Brett. Taurine transport in cultured choroid plexus. Pharm Res 14:406–409, 1997.

26. T Zeuthen, EM Wright. Epithelial potassium transport: Tracer and electrophysiological studies in choroid plexus. J Membr Biol 60:105–128, 1981.

27. A Hakvoort, M Haselbach, H-J Galla. Active transport properties of porcine choroid plexus cells in culture. Brain Res 795:247–256, 1998.

28. CE Johanson. Ontogeny and phylogeny of the blood–brain barrier. In: EA Neuwelt, ed. Implication of the Blood–Brain Barrier and Its Manipulation. Vol 1. New York: Plenum Press, 1986.

29. RH Angeletti, PM Novikoff, SR Juvvadi, JM Fritschy, PJ Meier, AW Wolkoff. The choroid plexus epithelium is the site of the organic anion transport protein in the brain. Neurobiology 94:283–286, 1997.

30. M Erben, S Decker, H Franke, H-J Galla. Electrical resistance measurements on cerebral capillary endothelial cells: A new technique to study small surfaces. J Biochem Biophys Methods 30:227–238, 1995.

31. J Wegener, M Sieber, H-J Galla. Impedance analysis of epithelial and endothelial cell monolayers cultured on gold surfaces. J Biochem Biophys Methods 32:151–170, 1996.

32. A Janshoff, J Wegener, M Sieber, H-J Galla. Double-mode impedance analysis of epithelial cell monolayer cultured on shear wave resonators. Eur Biophys J 25:93–103, 1996.

33. C Steinem, A Janshoff, WP Ulrich, W Willenbrink, M Sieber, J Wegener, H-J Galla. Impedance and shear wave resonance analysis of ligand–receptor interactions at functionalized surfaces and of cell monolayers. Biosensors Bioelectron 12:787–808, 1997.

34. J Wegener, A Janshoff, M Sieber, H-J Galla. Cell adhesion monitoring using a quartz crystal microbalance: Comparative analysis of different mammalian cell lines. Eur J Biophys 28:26–37, 1998.

35. J Wegener, S Zink, P Rösen, H-J Galla. Use of electrochemical impedance measurements to monitor β-adrenergic stimulation of bovine aortic endothelial cells. Eur J Physiol (in press, 1999).

4

In Vitro Models of the Blood–Brain Barrier and Their Use in Drug Development

R. Cecchelli, L. Fenart, and V. Buée-Scherrer
Université d'Artois, Lens, France

B. Dehouck, L. Descamps, C. Duhem, G. Torpier, and M. P. Dehouck
INSERM U325, Institut Pasteur, Lille, France

I. INTRODUCTION

Drug therapy of the central nervous system presents many logistical problems. The major problem, which is due to the presence of the blood–brain barrier (BBB), is drug delivery to the brain. Brain capillary endothelial cells forming the BBB are sealed by complex tight junctions and possess few pinocytotic vesicles. These characteristics, added to a metabolic barrier, restrict the passage of most small polar molecules and macromolecules from cerebrovascular circulation to the brain. The transfer is also controlled by a selective enzyme degradation at the endothelial border, in which P-glycoprotein actively extrudes various pharmaceutical molecules from the brain. As a result, a plethora of compounds that have demonstrated efficacy in vitro cannot be used as brain pharmaceutical agents in vivo.

Many endothelial functions that include diffusion to or transport from brain microvessels have been defined by studies in whole animals and in isolated capillaries in vitro. The ability to grow central nervous system microvascular endothelial cells in culture has opened the door to many new experimental approaches for studying the transendothelial transport of substances across the in vitro BBB.

Essentially three types of brain capillary endothelial cell culture are currently used by researchers: primary cultures, cell lines, and coculture systems. Each has demonstrated and contributed to basic information on cellular, biochemical, and molecular properties of the BBB (1–3).

The limitation of primary cultures has been their higher paracellular permeability, reflected by the measurement of the electrical resistance across the monolayer. This property apparently results from the development of incomplete tight junction in cell culture. Furthermore, brain microvessel endothelial cells undergo some dedifferentiation in cell culture, resulting in downregulation of the expression of some proteins, such as the glucose transporter (4). Since these solo cultures are free of other environmental aspects, brain microvessel endothelial cell function and differentiation in vitro may not be identical to those of the BBB in vivo.

Several transformed, immortalized endothelial cell systems that retain some BBB endothelial markers have been established. Different techniques have been used for prolonging the life span of cells in culture. This research led to the generation of rat, bovine, and human immortalized endothelial cells and their use as a replacement for primary cells in in vitro BBB models (5–7). However, these cell systems have not been characterized to the same extent as either primary or passaged cells (8).

In this chapter, we describe our in vitro model (8–11), which closely mimics the in vivo situation by culturing brain capillary endothelial cells on one side of a filter and astrocytes on the other. Moreover, it has already been widely used to study drug transport. This aspect is discussed, as well.

II. CULTURE OF BRAIN CAPILLARY ENDOTHELIAL CELLS

A. Brain Capillary Endothelial Cell Primary Cultures

Several procedures and many modifications have been used to obtain endothelial cell primary cultures. The species and source of cerebral microvessels vary from one laboratory to another, but usually all the procedures use mechanical means to disperse brain tissue (12). However, different approaches are used to collect and grow the disrupted capillaries.

The most critical points in the production of pure capillary endothelial cell cultures are the filtration and separation of microvessels from other brain constituents. Most of the laboratories use enzymatic digestion to isolate endothelial cells. The endothelial cell cultures prepared this way, from cerebral cortex and by using different sets of enzymes (collagenase and/or dispase), are in fact a mixture of cells of capillary, arteriolar, and venular origin. This primary culture is thus a heterogeneous mixture of these different cellular types (13).

The other way of culturing endothelial cells is the cloning of endothelial cell islands emerging from capillaries plated in vitro (10). We have been using this approach to obtain bovine brain capillary endothelial cells, without any help of enzymatic digestion. For this purpose, after isolation by mechanical homogenization from one cerebral hemisphere, microvessels consisting mainly of capillaries with a few tufts of arterioles and venules are seeded onto dishes coated with an extracellular matrix secreted by bovine corneal endothelial cells. Because only capillaries adhere to the extracellular matrix, arterioles and venules can easily be discarded. Five days after seeding, the first endothelial cells migrate out of the capillaries and start to form microcolonies. When the colonies are sufficiently large, the five largest islands are trypsinized and seeded onto 35 mm diameter, gelatin-coated dishes (one clone per dish) in the presence of Dulbecco's modified Eagle medium supplemented with 15% calf serum, 2 mM glutamine, 50 µg/ml gentamicin, 2.5 µg/ml amphotericin B, and basic fibroblast growth factor (bFGF, 1 ng/ml added every other day). Endothelial cells from one 35 mm diameter dish are then harvested and seeded onto 60 mm diameter gelatin-coated dishes. After 6–8 days, confluent cells are subcultured at the split ratio of 1:20. Cells at the third passage are stored in liquid nitrogen.

An obvious advantage of the use of cloned endothelial cells emerging from identified capillaries is that the culture is not contaminated by endothelial cells of arteriolar and venular origin. Moreover, pericytes and endothelial cells can be separated by microtrypsinization.

B. Subculture of the Brain Capillary Endothelial Cells

For experiments, cells are rapidly thawed at 37°C and seeded onto two 60 mm diameter gelatin-coated dishes. When they reach confluence, cells are subcultured up to passage 8. The life span of the endothelial cell cultures is about 50 cumulative population doublings. At each passage, the cells are seeded in stock gelatin-coated dishes and on microporous membranes.

The subculture technique enables us to circumvent the limitations of primary cultures and to provide large quantities of these monolayers. Indeed, endothelial cells can be cultured from passage 3 after thawing to passage 7, each passage generating at least 75 cocultures.

III.   COCULTURE OF BRAIN CAPILLARY ENDOTHELIAL
CELLS AND ASTROCYTES

The development of a cell culture system that mimics a biological barrier requires the establishment of the culture of endothelial cells on microporous supports. To reconstruct some of the in vivo complexities of the cellular environment, we developed this in vitro model of BBB by growing endothelial cells on one side of a filter and astrocytes on the bottom of 6-well plastic dishes. The culture medium is shared by both cell populations, allowing humoral interchange without any direct cell contact. Bovine brain capillary endothelial cells form a monolayer of small, tightly packed, nonoverlapping and contact-inhibited cells (9, 10). In these culture conditions, bovine brain capillary endothelial cells retain both endothelial (factor VIII–related antigen, angiotensin-converting enzyme) and the BBB features, including P-glycoprotein (14).

To verify that our model can be used for predicting drug transport across the BBB, we used our coculture system to compare results obtained in vivo and in vitro. However, the development of the cell culture system that mimics a specific BBB requires a microporous membrane covered with collagen. And for transport studies, this microporous membrane covered with collagen must be readily permeable to hydrophilic and hydrophobic solutes and to both low and high molecular weight molecules.

Before conducting transport studies with cell cultures on microporous membrane, it is essential that control experiments be performed using the microporous membrane coated with collagen only. These results will assure that the solute is freely permeable through the membrane and the collagen, and that the diffusion barrier is provided by the cell monolayer alone.

IV.   TRANSENDOTHELIAL TRANSPORT STUDIES

A.   Method

Transport studies are performed in the same way with cell cultures and with membranes coated with collagen only. On the day of the experiments, Ringer-HEPES (150 mM NaCl, 5.2 mM KCl, 2.2 mM $CaCl_2$, 0.2 mM $MgCl_2$, 6 mM $NaHCO_3$, 2.8 mM glucose, 5 mM HEPES) is added to the lower compartments of a 6-well plate (2.5 ml per well). One filter is then transferred into the first well of the 6-well plate containing Ringer, and Ringer-containing labeled or unlabeled drugs is placed in the upper compartment. At different times after addition of the labeled compound, the filter is transferred to another well of the 6-well plate to minimize the possible passage from the lower to the upper compartment. Incubations are performed on a rocking platform at 37°C. Indeed, shaking minimizes the thickness of the aqueous boundary layer on the cell monolayer surface, hence influences the permeability of lipophilic solutes.

An aliquot from each lower compartment and stock solution is taken, and the amount of drugs in each sample is measured by means of high performance liquid chromatography (HPLC) for unlabeled drugs, or in a liquid scintillation counter for labeled drugs.

Permeability calculations are performed as described by Siflinger-Birnboim et al. (15). To obtain a concentration-independent transport parameter, the clearance principle is used. The increment in cleared volume between successive sampling events is calculated by dividing the amount of solute transport during the interval by the donor chamber concentration. The total cleared volume at each time point is calculated by summing the incremental cleared volumes up to the given time point:

$$\text{Clearance (ml)} = \text{Cl (ml)} = \frac{X}{C_d}$$

where X is the amount of drug in the receptor chamber and C_d is the donor chamber concentration at each time-point.

During the 45-minute experiment, the clearance volume increases linearly with time. The average volume cleared is plotted versus time, and the slope is estimated by linear regression analysis. The slope of the clearance curves for the culture is denoted PS_t, where PS is the permeability $\times$ surface area product, in milliliters per minute. The slope of the clearance curve with the control filter coated only with collagen is denoted PS_f.

The PS value for the endothelial monolayer (PS_e) is calculated from

$$\frac{1}{PS_e} = \frac{1}{PS_t} - \frac{1}{PS_f}$$

The PS_e values are divided by the surface area of the porous membrane to generate P_e, the endothelial permeability coefficient, in centimeters per minute.

B. Drug Transport Across the Blood–Brain Barrier: Correlation Between In Vitro and In Vivo Models

The oil–water partition coefficients of compounds with different lipid solubilities have often been claimed to give fairly good predictions of the BBB penetration. When the brain uptake index (BUI) values are plotted versus the partition coefficients (olive oil–water) reported by Oldendorf for the same set of compounds as above, however, a poor correlation ($r = 0.51$) results, as can be seen in Fig. 1.

One of our major objectives was to ascertain, by means of an in vitro model of the BBB, whether it is possible to predict the in vivo BBB permeability and brain uptake of a number of compounds having different physicochemical properties and able to penetrate the BBB via passive diffusion. This determination was assessed by making in vitro/in vivo comparisons for well-known molecules. Figure 2 plots the log natural of the brain uptake index for nine compounds versus the log natural of the in vitro P_e values. There is in this case a good correlation ($r = 0.93$, $p < 0.0008$) between the in vitro and in vivo data, except for imipramine (not used in the regression analysis). Although imipramine is a very lipophilic drug with a high BUI value, its in vitro P_e value turned out to be low. This low P_e value was found to be due to substantial sequestration of the compound within the endothelial cells in vitro. One explanation for the high BUI value could then be that since a similar differentiation of endothelial binding/endocytosis from actual transcytosis was not performed in vivo [this requires the BUI technique to be complemented with a capillary depletion technique (16)], the BUI value for imipramine could not reflect the actual transport of the compound through the BBB in vivo.

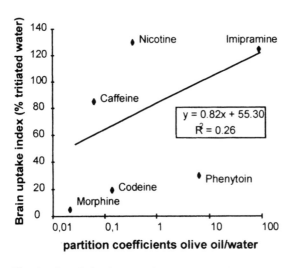

Fig. 1 Correlation between the partition coefficients of olive oil in water and in vivo brain uptake index.

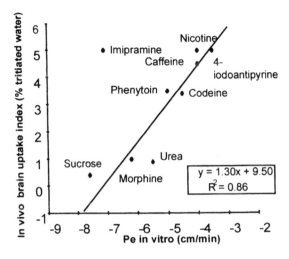

Fig. 2 Correlation between in vivo brain uptake index and in vitro permeability (Pe) of the blood–brain barrier model.

C. Screening of Analogues

The relative ease with which the coculture can be produced in large quantities, and the strong correlations obtained between the in vitro and in vivo results, suggest that the coculture is an efficient system for the screening of centrally active drugs. We screened a series of 92 different analogues in 23 experiments using 10 different colonies of endothelial cells. The results presented in Fig. 3 demonstrate the reproducibility of the model for the control molecule: the permeability of endothelial cell monolayers for the different analogues is compared to the permeability for the original molecule (control molecule). This study allows one to determine, in a very short time, whether the chemical modification of the control molecule can or cannot increase its transport across the BBB. The analyses were performed by HPLC.

V. USE OF ENDOGENOUS ROUTES

From the viewpoint of drug delivery, it is important to understand and try to use the specific transport mechanisms of brain capillary endothelial cells. Indeed transcytosis of the antibody against the ferrotransferrin receptor (OX-26) has already been reported by Pardridge et al. (17), who demonstrated that the OX-26 selectively targets brain relative to organs such as kidney, heart,

PSt/PSf

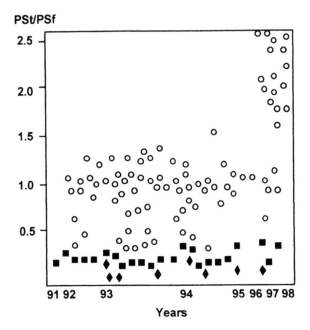

Fig. 3 Screening of drugs. Squares, control molecule; circles, Pe ≥ control molecule; diamonds, Pe ≤ control molecule.

or lung. This statement is supported by the capillary depletion experiments of Friden et al. (18, 19), who demonstrated the blood-to-brain entry of OX-26 conjugates to methotrexate and nerve growth factor, and by Broadwell et al. (20), who used horseradish peroxidase–conjugated OX-26 to investigate cyto-chemical approaches. Since the neurological abnormalities that result from the inadequate absorption of dietary vitamin E can be improved by the oral administration of pharmacological doses of vitamin E, Traber and Kayden (21) have suggested that low density lipoprotein (LDL) functions as a transport system for tocopherol to the brain. Moreover, studies with lipoproteins have demonstrated that the entire fraction of the drug bound to LDL is available for entry into the brain.

With the use of our coculture of bovine brain capillary endothelial cells and astrocytes, we provided evidence for the selective transport of LDL and iron-loaded transferrin (Tf) (22–24), across the endothelial cell monolayers. Both transports were completely inhibited at low temperature, indicating that these blood-borne molecules are directed to the abluminal compartment by a transcellular route. Moreover, when the conventional antireceptor antibodies (OX 26 and C7) were used, a complete inhibition of Tf and LDL endothelial cell uptakes was observed. Taken together, these results led us to conclude that Tf and LDL are transcytosed through the BBB by way of a specific receptor-mediated pathway.

Interestingly, no intraendothelial degradation of LDL or Tf was observed, indicating that the transcytotic pathway in brain capillary endothelial cells is different from these receptor classical pathways, which involve clathrin-coated pits, coated vesicles, endosomes, and lysosomes. Although clathrin-coated vesicles have been the most extensively studied, there are various other clathrin-independent plasmalemmal vesicles that may also function in the trafficking of molecules at cell surfaces. Caveolae are one distinctive type of non-clathrin-coated plasmalemmal vesicles. More recent work suggests that caveolae mediate not only fluid phase but also receptor-mediated endocytosis or transcytosis of molecules.

Our studies demonstrate that the transcytotic LDL and Tf pathways in brain capillary endothelial cells are different from the classical pathways of these substances and suggest that a caveolin–endocytic pathway is involved in their uptake and traffic through the endothelial cells.

The highly differentiated, reproducible in vitro model of brain endothelial cells in coculture with glial cells will allow us to gain insight into the mechanism that regulates transcellular transport and will lead to better delivery systems. Using these endogenous routes, we are able to deliver to the brain a large amount of albumin loaded in specific synthetic carriers.

VI. CONCLUSIONS

The in vitro BBB model, consisting of a coculture of brain capillary endothelial cells on one side of a filter and astrocytes on the other, is currently used by pharmaceutical companies. The strong correlation between the in vivo and in vitro values demonstrated that this in vitro system is an important tool for the investigation of the role of the BBB in the delivery of nutrients or drugs to the central nervous system. The main advantage of this model is the possible rapid evaluation of strategies for achieving drug targeting to the CNS or to appreciate the eventual central toxicity of systemic drug.

The relative ease with which such cocultures can be produced in large quantities offers advantages over conventional techniques including rapid assessment of the potential permeability of a drug, and the opportunity to elucidate the molecular transport mechanism of substances across the BBB.

REFERENCES

1. F Joo. The blood–brain barrier in vitro; ten years of research, on microvessels isolated from the brain. Neurochem Int 7:1–25, 1985.
2. F Joo. The cerebral microvessels in culture, an update. J Neurochem 58:1–17, 1992.
3. F Joo. The blood–brain barrier in vitro: The second decade. Neurochem Int 23: 499–521, 1993.
4. RJ Boado, WM Pardridge. Molecular cloning of the bovine blood–brain barrier glucose transporter cDNA and demonstration of phylogenetic conservation of the 5′-untranslated region. Mol Cell Neurosci 1:224–232, 1990.
5. CA Diglio, DE Wolfe, P Meyers. Transformation of rat cerebral endothelial cells by Rous sarcoma virus. J Cell Biol 97:15–21, 1983.
6. F Roux, O Durieu-Trautmann, N Chaverot, M Claire, P Mailly, JM Bourre, AD Strosberg, PO Couraud. Regulation of gamma-glutamyl transpeptidase and alkaline phosphatase activities in immortalized rat brain microvessel endothelial cells. J Cell Physiol 159:101–113, 1994.
7. RD Hurst, IB Fritz. Properties of an immortalised vascular endothelial/glioma cell coculture model of the blood–brain barrier. J Cell Physiol 167:81–88, 1996.
8. S Méresse, MP Dehouck, P Delorme, M Bensaïd, JP Tauber, C Delbart, JC Fruchart, R Cecchelli. Bovine brain endothelial cells express tight junctions and monoamine oxidase activity in long-term culture. J Neurochem 53:1363–1371, 1989.
9. MP Dehouck, S Méresse, P Delorme, JC Fruchart, R Cecchelli. An easier, reproductible and mass production method to study the blood–brain barrier ''in vitro'' and ''in vivo'' models. J Neurochem 54:1790–1797, 1990.
10. MP Dehouck, P Jolliet-Riant, F Brée, JC Fruchart, R Cecchelli, JP Tillement. Drug transfer across the blood–brain barrier: Correlation between in vitro and in vivo models. J Neurochem 58:1790–1797, 1992.
11. MP Dehouck, B Dehouck, C Schluep, M Lemaire, R Cecchelli. Drug transport to the brain: Comparison between in vitro and in vivo models of the blood–brain barrier. Eur J Pharm Sci 3:357–365, 1995.
12. AG De Boer, W Sutanto. Perspectives. In: AG de Boer, W Sutanto, eds. Drug Transport Across the Blood–Brain Barrier. Amsterdam: Harwood Academic Publishers, 1997, pp 215–216.
13. B Dehouck, R Cecchelli, MP Dehouck. A co-culture of brain capillary endothelial cells and astrocytes: An in vitro blood–brain barrier for studying drug transport to the brain. In: AG de Boer, W Sutanto, eds. Drug Transport Across the Blood–Brain Barrier. Amsterdam: Harwood Academic Publishers, 1997, pp 69–79.
14. L Fenart, V Buée-Scherrer, L Descamps, C Duhem, MG Poullain, R Cecchelli, MP Dehouck. Inhibition of P-glycoprotein: Rapid assessment of its implication

in blood–brain barrier integrity and drug transport to the brain by an in vitro model of the blood–brain barrier. Pharm Res 15:993–1000, 1998.

15. A Siflinger-Birnboim, PJ Del Becchio, JA Cooper, FA Blumenstock, JN Shepard, AB Malik. Molecular sieving characteristics of the cultured endothelial monolayer. J Cell Physiol 132:111–117, 1987.

16. D Triguero, JB Buciak, WM Pardridge. Capillary depletion method for quantification of blood–brain barrier transport of circulating peptides and plasma proteins. J Neurochem 54:1882–1888, 1990.

17. WM Pardridge, J Eisenberg, WT Cephalu. Absence of albumin receptor on brain capillaries in vivo and in vitro. Am J Physiol 249:E264–E267, 1985.

18. PM Friden, LR Walus, GF Musso, MA Taylor, B Malfroy, RM Starzyk. Anti-transferrin receptor antibody and antibody–drug conjugates cross the blood–brain barrier. Proc Natl Acad Sci USA 88:4771–4775, 1991.

19. PM Friden, LR Walus, P Watson, SR Doctrow, JW Kozarich, C Backman, H Bergman, B Hoffer, F Bloom, AC Granholm. Blood–brain barrier penetration and in vivo activity of an NGF conjugate. Science 259:373–377, 1993.

20. RD Broadwell, BJ Balin, M Saleman. Transcytotic pathway for blood-borne protein through the blood–brain barrier. Proc Natl Acad Sci USA 85:632–636, 1988.

21. MG Traber, HJ Kayden. Vitamin E is delivered to cells via the high affinity receptor for low density lipoprotein. Am J Clin Nutr 40:747–751, 1984.

22. B Dehouck, MP Dehouck, JC Fruchart, R Cecchelli. Upregulation of the low density lipoprotein receptor at the blood–brain barrier: Intercommunications between brain capillary endothelial cells and astrocytes. J Cell Biol 126:465–473, 1994.

23. B Dehouck, L Fenart, MP Dehouck, A Pierce, G Torpier, R Cecchelli. A new function for the LDL receptor: Transcytosis of LDL across the blood–brain barrier. J Cell Biol 138:877–888, 1997.

24. L Descamps, MP Dehouck, G Torpier, R Cecchelli. Receptor-mediated transcytosis of transferrin through blood–brain barrier endothelial cells. Am J Physiol 270:H1150–H1158, 1996.

5

The Application of Microdialysis Techniques to the Study of Drug Transport Across the Blood–Brain Barrier

A. G. de Boer, Elizabeth C. M. de Lange, Inez C. J. van der Sandt, and Douwe D. Breimer
Leiden/Amsterdam Center for Drug Research, University of Leiden, Leiden, The Netherlands

I. INTRODUCTION

Many brain diseases (e.g., epilepsy, Parkinson's disease, Alzheimer's disease, schizophrenia, depression, ischemia, edema) arise from local or peripheral physiological disorders. Others (e.g., encephalitis, meningitis, AIDS dementia) are caused by brain infections. Therefore drugs that are effective against diseases in the central nervous system (CNS) and reach the brain via the blood compartment must pass the blood–brain barrier (BBB). This barrier is considered to be the most important barrier for drug transport to the brain because its surface area is 5000 times larger than the blood–cerebrospinal fluid barrier located at the choroid plexuses.

The BBB is presented by the endothelial lineage of the capillaries in the brain and has special properties that distinguish it from peripheral endothelium (1–3). In particular, it limits the access of hydrophilic compounds to the brain, and several import and export systems are present (4). In addition to its physical barrier properties, the BBB must be considered to be a metabolic barrier (5), while immunological barrier properties are displayed by the endothelial cells together with surface molecules, which play a key role in such pathological conditions as inflammation, tumor angiogenesis, and wound healing (6).

The function of the BBB is dynamically regulated by various cells present at the level of the BBB (Fig. 1) (3), comprising astrocytes, neurons, pericytes, and microglia, but also cells (leukocytes) from the general circulation including peripheral and local hormones (cytokines). Thus the BBB is a very complex and tight system (2, 7). Because of the fenestrated endothelium, however, the BBB is physically leaky in various parts of the brain. In particular, in circumventricular organs (CVOs) such as the area postrema or the median eminence (8), the endothelium is leaky, while in the choroid plexus, the choroid epithelium induces the formation of fenestrated capillaries that are highly permeable to intravenously injected horseradish peroxidase (9).

Apart from passive hydrophilic paracellular (tight junctional) and passive lipophilic transcellular transport, there are several other possibilities for transcellular drug transport across the BBB. These include fluid phase and adsorptive-mediated transcytosis, while carriers and transporters are involved, as well. Several of these transporter systems have been studied—for example, the acidic amino acid system, the peptide transporter, the glucose transporter (GLUT-1), the P-glycoprotein, and the amino acid transport systems: the A

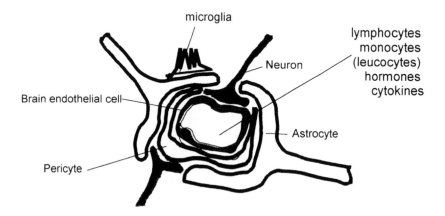

Fig. 1   Cell types, including peripheral and local hormones, present at and influencing the functionality of brain capillary endothelial cells comprising the blood–brain barrier (BBB). (Modified from Ref. 3.)

system (glycine and proline; situated abluminally and transports out of the brain), the large neutral amino acid, the L system (tyrosine, phenylalanine, leucine, isoleucine, valine, histidine, and methionine; situated luminally and abluminally and transport into the brain), and the ASC system (alanine, serine, cystine, threonine, asparagine; situated luminally and abluminally and transports into the brain). The discovery of the presence of P-glycoprotein (Pgp) at the blood–brain barrier (10, 11) has contributed an especially valuable part of our understanding of the penetration of various drugs into the brain. A 170 kDa glycoprotein, Pgp belongs to the superfamily of the ATP-binding cassette (ABC) transporters (12); it is involved in particular in the efflux of (cytostatic) drugs from the endothelial compartment and therefore from the brain into blood again and is responsible for the appearance of multidrug resistance (13, 14).

To understand the pharmacokinetics and pharmacodynamics of drugs in the CNS, it is important to know their unbound concentration in the extracellular fluid of the brain. Various techniques are available to study this property in vitro (15) and in vivo (16). The in vivo techniques include the brain uptake index (BUI) (17), the brain efflux index (BEI) (18), brain perfusion (19, 20), the unit impulse response method (21, 22), and microdialysis (16).

Microdialysis is our method of choice in the study in of vivo drug transport across the BBB for the following reasons. Based on physiological and anatomical considerations, the brain cannot be considered to be a homogeneous compartment. In addition, drug disposition in the brain is determined by protein binding, blood flow, BBB transport, and the exchange between brain extracellular fluid (ECF) and brain cells. Intracerebral microdialysis is particularly suitable for estimating (local) extracellular unbound drug concentrations as a function of time. When the kinetics of the unbound drug in blood is known, intracerebral microdialysis can be used to characterize drug transport across the BBB under various (disease) conditions. In addition, since unbound concentrations are measured by microdialysis, a correlation can be made between the pharmacokinetics of the drug in blood and brain and its pharmacodynamic effect(s) at the brain, which one assumes to be related to the unbound concentration of the drug in the ECF.

The most important advantage of the microdialysis technique is that it provides samples obtained at multiple time points from an individual (freely moving) animal. This may lead to a significant reduction of the number of animals needed to determine pharmacokinetic profiles of unbound drug in the ECF of the brain. Since the microdialysis probe can be implanted in virtually any brain region, providing local information on the drug concentration at the selected site, pharmacokinetic profiles can be obtained in normal brain as well as in affected or diseased brain sites, for example, in brain tumors. Nevertheless, intracerebral microdialysis is an invasive technique: it involves the implantation of a probe, which may cause tissue trauma, hence may have consequences for BBB function (23–29). Therefore it was necessary to determine whether intracerebral microdialysis provides meaningful data on drug transport across the BBB and drug disposition in the brain. A critical discussion on the methodological considerations of the application of microdialysis in this respect has been given elsewhere (16). In addition, a comparison with other in vivo methods was made (26, 28). The paragraphs that follow briefly describe the technical aspects of microdialysis.

Intracerebral microdialysis involves the stereotactic implantation of a microdialysis probe in the brain. The probe comprises a semipermeable membrane, partly covered by an impermeable coating. The probe may be positioned at a specific site in the brain and perfused with an artificial ECF solution. It may be used in the delivery mode but also in the sampling mode. The latter is used most frequently and is supposed to dialyze compounds from the ECF behind the BBB into the probe. Subsequently, the dialysate is collected and assayed by various online or offline techniques (30, 31).

Although the principle is simple, data interpretation may be more complicated, owing to the complexity of the relation between brain extracellular fluid and dialysate concentration (in vivo concentration recovery). This relationship depends not only on probe characteristics (in vitro concentration recovery), but also on such periprobe processes as generation, elimination, metabolism, and intra-/extracellular exchange of the compound (32, 33). Therefore, in vivo concentration recovery should be assessed for each experiment (34–39).

II. EXAMPLES

Some examples are given to illustrate the applicability of microdialysis to the study of drug transport to the brain. A study on osmolar brain opening, drug penetration into a brain tumor, and the effects of P-glycoprotein on brain distribution of rhodamine-123 (R123) is presented to supply more detail.

A. Relation Between BBB Transport and Lipophilicity

It was of interest to determine whether the correlation between lipophilicity and extent of BBB transport is reflected also by the microdialysis technique.

Optimal experimental conditions were carefully selected (26, 40). Atenolol and acetaminophen were used as model drugs, representing a hydrophilic drug (log $P = -1.78$) and a moderately lipophilic drug (log $P = 0.25$), respectively. The ratio of the area under the curve of brain$_{ECF}$ over that in plasma ($AUC_{brain\ ECF}/AUC_{plasma}$) was used as a measure of BBB transport. This ratio was 18% for acetaminophen, while for atenolol it was only 4% (Fig. 2) (26). Applications of other techniques show similar results; that is, lipophilic drugs crossed the BBB more readily than hydrophilic ones. Moreover, these observations were confirmed by using a parallel microdialysis probe design to measure the spatial distribution of these drugs after local brain infusion (41).

B. BBB Opening

Another approach to the study of BBB transport characteristics was to investigate the effect of osmotic BBB opening together with intravenous administration of atenolol. Since atenolol is poorly transported across the BBB, transport of this drug to the brain was expected to increase considerably under these conditions. A hypertonic mannitol solution was infused into the left carotid artery, and atenolol concentration–time profiles were measured at the ipsilat-

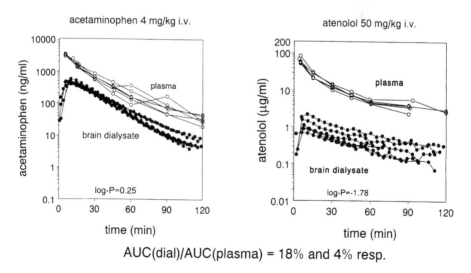

Fig. 2 Brain dialysate and plasma concentrations following intravenous administration of 825 µg acetaminophen (left) and 10 mg atenolol (right). (Redrawn from Ref. 26.)

eral and contralateral hemisphere, as well as after the infusion of saline at the ipsilateral side. An approximately 10-fold increase in atenolol $AUC_{dialysate}$ was found only at the ipsilateral hemisphere of hyperosmolar mannitol infusion, clearly reflecting BBB opening (Fig. 3) (42). Interestingly, a more detailed evaluation of the data indicated that a circadian variation in the effect of hyperosmotic mannitol infusion on the transport of atenolol exists: only afternoon experiments resulted in this increase of BBB permeability (native BBB transport of atenolol did not show time-of-day variability).

C. Distribution of Methotrexate in Normal and Tumor-Bearing Brain

Methotrexate is a hydrophilic anticancer agent ($\log P = -1.85$) and is poorly transported into the brain, probably by restricted paracellular BBB transport. Because intracerebral microdialysis reflects local $brain_{ECF}$ concentrations (12), it was applied to investigate possible changes in the distribution of methotrexate in tumor-bearing brain. Brain penetration was determined in normal brain,

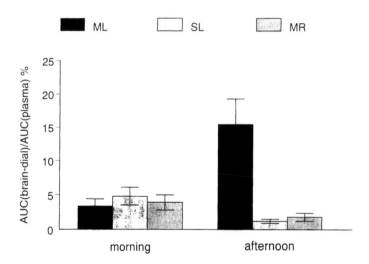

Fig. 3 Ratio of $AUC_{brain\text{-}dialysate}/AUC_{plasma}$ after intravenous administration of 10 mg atenolol as obtained in the left (ML) or right (MR) brain cortex following infusion of hyperosmolar mannitol into the left internal carotid artery, or in the left cortex after saline infusion into the left internal carotid artery (SL): difference between morning and afternoon experiments. (Redrawn from Ref. 42.)

in brain ipsilateral to sham tumor implantation, and in brain ipsilateral as well as contralateral to tumor (rhabdomyosarcoma) implantation (Fig. 4) (43). Penetration was about 5% in normal and sham implanted brain. The presence of tumor tissue increased the penetration of methotrexate into the ipsilateral brain by approximately 150%, whereas at the contralateral cortex the BBB permeability seemed to be reduced (to 65%). The change in methotrexate penetration could be compared to methotrexate uptake data of earlier studies in which similar values were observed.

D. Active BBB Transport as Estimated in *mdr1a* (−/−) and (+/+) Mice

Pgp is found in many types of cancer tissue but is also present in cells with a barrier function, like the endothelial cells in the brain capillaries (10, 11, 44, 45) comprising the BBB. Pgp is expressed on the luminal side of these cells and may therefore counteract brain penetration of Pgp substrates. Strong evidence for this is provided by the large differences observed in BBB transport for a number of Pgp substrates between *mdr1a* (−/−) and wild-type

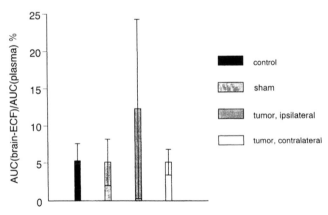

Fig. 4 Individual ratios of $AUC_{brain-dialysate}/AUC_{plasma}$ after intravenous bolus administration of methotrexate (75 mg/kg) to control, sham surgery with ipsilateral measurements, tumor implantation and ipsilateral measurement, and tumor implantation and contralateral measurement. (Redrawn from Ref. 43.)

$(+/+)$ mice (13, 46) based on brain homogenates and studies with intracerebral microdialysis.

Rhodamine-123 (R123), which is a fluorescent compound and widely used to study Pgp functionality, was taken as a model substrate (47). Brain homogenate studies showed that blood (plasma) levels were not different between $mdr1a$ $(-/-)$ and $(+/+)$ mice, while brain levels were about four-fold higher in $mdr1a$ $(-/-)$ mice, clearly indicating higher BBB transport in the latter. In addition, no significant differences were found in R123 concentration in different parts of the brain.

Subsequently, the potential influence of various experimental conditions on Pgp function was investigated following the introduction of a microdialysis probe and surgery. No significant changes in total brain and blood concentrations between the various surgical and experimental conditions were observed in $mdr1a$ $(-/-)$ or $(+/+)$ mice.

It was concluded that microdialysis could be reliably used to study in vivo Pgp function at the level of the BBB, excluding artifacts in the interpretation of Pgp function introduced by the intracerebral microdialysis methodology.

The difference in R123 brain homogenate concentrations between $mdr1a$ $(-/-)$ and $(+/+)$ mice, however, was not observed to the same degree in dialysate concentrations from these mice (48). In a subsequent study, it was investigated whether this could be due to differences between in vivo recovery of R123 in $mdr1a$ $(-/-)$ and $(+/+)$ mice. Therefore, in vivo concentration recovery was determined by the no net flux (NNF) method (49). Indeed, it could be shown that in vivo recovery of the microdialysis probe was different. The recovery values were $4.8 \pm 6.2\%$ and $17 \pm 7.2\%$ for the $mdr1a$ $(-/-)$ and $mdr1a$-$(+/+)$ mice, respectively. Correcting the ECF values of R123 obtained following intracerebral microdialysis revealed ECF concentrations of 15.7 ± 7.8 nM and 3.6 ± 2.1 nM for the $mdr1a$ $(-/-)$ and $(+/+)$ mice, respectively, showing that accumulation in $mdr1a$ $(-/-)$ mice had increased by a factor of 4.

III. DISCUSSION

The first example illustrated BBB transport by passive lipophilic diffusion. In these experiments atenolol and acetaminophen were given intravenously to rats, and brain extracellular fluid (ECF) concentrations were estimated by microdialysis (26); for concentration–time profiles measured in the ECF and in plasma, see Fig. 2. It can be seen that the ECF concentrations follow very nicely the plasma concentrations and that the ECF concentrations of atenolol were much smaller than those of acetaminophen. This can largely be explained by their difference in log P values [log P (atenolol) = -1.78; log P (acetaminophen) = 0.25] and subsequently their differences in lipophilic diffusion across the BBB.

The second example illustrates the estimation of BBB opening by a hypertonic mannitol solution administered in the left carotid artery of rats (42). Figure 3 showed that atenolol concentrations were increased in brain ECF compared to control saline administration in the same or to mannitol administration in the other brain hemisphere. Here a time-dependent opening of the BBB was observed, indicating a difference between morning and afternoon experiments. One may conclude that rats were less vulnerable to BBB opening in the morning than in the afternoon. The interval between probe implantation and start of the experiment did not appear to be the cause of this. In addition, it is conceivable that not all the infused volume is equally distributed over the left hemisphere during the day, so that local threshold levels needed to open the BBB may not have been achieved in the morning. It may be also because of changed blood flow. Alternatively, differences in sensitivity of regional brain tissue to the hyperosmotic insult (intrinsically) or by interaction of endogenous compounds that show circadian variation in their levels within the periprobe tissue could also have caused this circadian effect. There seems to be an inverse correlation with the height of corticosterone levels in plasma, since it has been shown that corticosterone levels increase considerably at about 15.00 hours (50). It may be speculated that although corticosteroids are thought to protect tissue against injury, the regulation of cellular homeostasis and the recovery following injury may be inhibited as well. Changes in heart rate and blood pressure could also explain the increased effects in afternoon experiments (51). However these parameters were not measured. Further studies are needed to reveal the reason for these effects. Nevertheless, it stresses the need for time-of-day standardization in BBB transport experiments, particularly following hyperosmolar manipulation.

The third example, performed following intravenous administration of methotrexate to rats and following implantation of a rhabdomyosarcoma in the brain of rats, illustrates the possibility of measuring the concentrations of cytostatic drugs in brain ECF and in brain tumors by microdialysis (see plasma and ECF concentrations of methotrexate in Fig. 4) (43). In addition, methotrexate concentrations could be measured in implanted tumors in rat brain.

Concentrations of methotrexate in tumors were tending to be higher than in brain ECF of control animals. This indicates that the BBB may be less tight or functional in such animals than in those having healthy brain tissue. Histological evaluation of the brains used in the microdialysis studies revealed the position of the probe in a tumor, if present. This could be in the core of the tumor, as well as in more diffuse areas, or even outside the tumor. It appeared that the presence of tumor tissue affected the concentrations of methotrexate in the brain. In few cases hyaline proteinaceous exudate was present, which was accompanied by increased penetration of methotrexate into the brain. This is in line with the fact that hyaline proteinaceous exudate is associated with increased vascular permeability. Other parameters, like vacuolization, hypercellularity, infiltration of granulocytes, hemorrhages, fibrosis, or necrosis did not seem to correlate with the pharmacokinetic data. These results indicate that intracerebral microdialysis can be used to investigate the effects of a brain tumor as well as other parameters on local BBB permeability (41, 43).

The last example illustrates the application of microdialysis in the estimation of BBB transport in *mdr1a*–Pgp-deficient and ($+/+$) mice. These mice were generated by Schinkel et al. (14). We have used rhodamine-123 as a Pgp substrate to determine its transport into the brain of *mdr1a*-deficient mice (48). Concentrations in brain homogenates of *mdr1a* ($-/-$) mice were four times higher than in *mdr1a* ($+/+$) mice. However, microdialysis experiments showed no substantial difference between brain ECF concentrations of R123 in *mdr1a* ($-/-$) and ($+/+$) mice. The decreased recovery can be explained by the Pgp deficiency in the *mdr1a* ($-/-$) mice, which leads to a higher resistance for mass transport from the endothelial cells to the ECF and therefore results in a smaller in vivo recovery value. This is substantiated and predicted by a theory developed by Bungay et al. (49). These researchers have theoretically shown that when the resistance against mass transfer (R_e) increases, the recovery also decreases. Figure 5 shows that the ratio $R_{e(-/-)}/R_{e(+/+)}$ in brain tissue is larger than one, and therefore $R_{e(-/-)}$ is larger than $R_{e(+/+)}$. This means that the mass transfer resistance in the ($-/-$) mice is larger than the mass transfer resistance in the ($+/+$) mice. A larger value for R_e will lead to a smaller in vivo recovery. This correlates well with the microdialysis data obtained in *mdr1a* ($-/-$) and ($+/+$) mice (48, 52).

A complication in the evaluation of these data is the observation that R123 is not a specific substrate for Pgp but is also transported by the organic cation carrier (53). It is not known whether this transporter is also present at the BBB. In vitro data obtained in MDR1-transfected LLC-PK1 (pig kidney

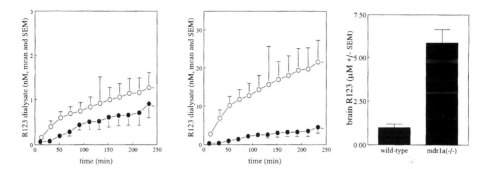

Fig. 5 Concentrations of rhodamine-123 (R123): uncorrected dialysate concentrations (left), dialysate concentrations corrected for in vivo recovery (middle), and brain homogenate concentrations (right) in *mdr1a* ($-/-$) and ($+/+$) mice. (From Ref. 48.)

tubular epithelial cell line) cells showed that transport of R123 could be inhibited by selective inhibitors of the organic cation transporter (53, 54).

IV. DISCUSSION AND CONCLUSIONS

The intracerebral microdialysis technique has improved considerably during the last decade. Today, this technique may be considered to be a useful tool for investigating the transport of drugs across the BBB into the brain. Various examples illustrated the applicability of microdialysis to study drug transport to the brain under various (disease) conditions.

In vivo concentration recovery is a very important issue in each experiment. For drugs that are passively transported across the BBB and are not metabolized in the brain to a large extent, changes in drug transport across the BBB will mainly involve changes in the influx of the drug into the brain. This is not likely to be accompanied by changes in in vivo concentration recovery, meaning that a relative comparison of BBB transport data of such drugs is valid.

However, for drugs whose transport is influenced by active influx/efflux transport systems, in vivo concentration recovery is dependent on these systems. This has been exemplified by the difference in in vivo concentration recovery between *mdr1a* ($-/-$) and ($+/+$) mice, where a change in active transport out of the brain was shown to have an influence on in vivo concentration recovery.

Additional problems may be elicited by dialyzing very lipophilic drugs (adhesion to tubing material). Another signal is limited sensitivity of HPLC detectors when very low concentrations of drugs are measured in small volumes of dialysate as in mice. In these cases one may switch to offline detection to permit the use of more sensitive methods (e.g., mass spectrometry).

Thus under carefully controlled experimental conditions, the use of intracerebral microdialysis appears to be successful in monitoring concentrations of drugs in a selected area of the brain. Combined with plasma concentration measurements, BBB transport characteristics can be determined in vivo. The technique offers the possibility of investigating BBB function under physiological and pathological conditions.

REFERENCES

1. JBMM van Bree, AG de Boer, M Danhof, DD Breimer. Drug transport across the blood–brain barrier. I. Anatomical and physiological aspects. Pharm Weekbl Sci Ed 14(5):305–310, 1992.
2. AG de Boer, DD Breimer. The blood–brain barrier (BBB): Clinical implications for drug delivery to the brain. J R Soc Physicians London 28(6):1–9, 1994.
3. WM Pardridge. Peptide Drug Delivery to the Brain. Raven Press, New York, 1991.
4. DJ Begley. The blood–brain barrier: Principles for targeting peptides and drugs to the central nervous system. J Pharm Pharmacol 48:136–146, 1995.
5. A Minn, JF Ghersi-Egea, R Perrin, B Leininger, G Siest. Drug metabolizing enzyme in the brain and cerebral microvessels. Brain Res Rev 16:65–82, 1991.
6. HG Augustin, DH Kozian, RC Johnson. Differentiation of endothelial cells: Analysis of the constitutive and activated endothelial cell phenotypes. BioEssays 16(12):901–906, 1994.
7. AG de Boer, DD Breimer. Cytokines and blood–brain barrier permeability. In: HS Sharma, J Westman, eds. Progress in Brain Research. Vol 115. Amsterdam: Elsevier, 1998, pp 425–451.
8. PM Gross, NM Sposito, SE Pettersen, JD Fenstermacher. Differences in function and structure of the capillary endothelium in grey matter, white matter and a circumventricular organ of rat brain. Blood Vessels 23:261–270, 1986.
9. J Wilting, B Christ. An experimental and ultrastructural study on the development of the avian choroid plexus. Cell Tissue Res 255:487–494, 1989.
10. C Cordon-Cardo, JP O'Brien, D Casals, L Rittman-Grauer, JL Bieler, MR Melamed, JR Bertino. Multidrug-resistance gene (P-glycoprotein) is expressed by

endothelial cells at blood–brain barrier sites. Proc Natl Acad Sci USA 86:695–698, 1989.

11. F Thiebaut, T Tsuruo, H Hamada, MM Gottesman, I Pastan, MC Willingham. Cellular localization of the multidrug resistance gene product in normal human tissues. Proc Natl Acad Sci USA 84:7735–7738, 1987.

12. CF Higgins, ID Hiles, GPC Salmonel, DR Gill, JA Downie, IJ Evans, IB Holland, L Gray, SD Buckel, AW Bell, MA Hermodson. A family of related ATP-binding subunits coupled to many distinct biological processes in bacteria. Nature 323:448–450, 1986.

13. AH Schinkel, JJM Smit, O Van Tellingen, JH Beijnen, E Wagenaar, L Van Deemter, CAAM Mol, MA Van der Valk, EC Robanus-Maandag, HPJ te Riele, AJM Berns, P Borst. Disruption of the mouse *mdr1a* P-glycoprotein gene leads to a deficiency in the blood–brain barrier and to increased sensitivity to drugs. Cell 77:491–502, 1994.

14. P Borst, AH Schinkel, JJM Smit, E Wagenaar, L Van Deemter, AJ Smith, EWHM Eijdems, F Baas, GJR Zaman. Classical and novel forms of *MDR*, and the physiological functions of P-glycoproteins in mammals. Pharmacol Ther 60:289–299, 1993.

15. AG de Boer, DD Breimer. Reconstitution of the blood–brain barrier in cell culture for studies of drug transport and metabolism. Adv Drug Delivery Rev 22:251–264, 1996.

16. ECM de Lange, AG de Boer, DD Breimer. Monitoring in vivo BBB drug transport: CSF sampling, the unit impulse response method and, with special reference, intracerebral microdialysis, STP Pharma Sci 7(1):17–28, 1997.

17. WH Oldendorf. Measurement of brain uptake of radiolabelled substances using a tritiated water internal standard. Brain Res 24:372, 1970.

18. A Kakke, T Terasaki, Y Sugiyama. Brain efflux index as a novel method of analyzing efflux transport at the blood–brain barrier. J Pharmacol Exp Ther 277:1550–1559, 1996.

19. BV Zlokovic, DJ Begley, BM Djuricic, DM Mitrovic. Measurement of solute transport across the blood–brain barrier in the perfused guinea pig brain: Method and application to *N*-methyl-α-aminoisobutyric acid. J Neurochem 46:1444–1451, 1986.

20. DJ Begley, LK Squires, BV Zlokovic, DM Mitrovic, CCW Hughes, PA Revest, J Greenwood. Permeability of the blood–brain barrier to the immunosuppressive cyclic peptide cyclosporin A. J Neurochem 55:1222–1230, 1998.

21. JBMM van Bree, AV Baljet, A van Geyt, AG de Boer, M Danhof, DD Breimer. The unit impulse response procedure for the pharmacokinetic evaluation of drug entry into the central nervous system. J Pharmacokinet Biopharm 17:441–462, 1989.

22. U Jaehde, MWE Langemeijer, AG de Boer, DD Breimer. Cerebrospinal fluid transport and disposition of the quinolones ciprofloxacin and pefloxacin. J Pharmacol Exp Ther 263(3):1140–1146, 1992.

23. H Benveniste, J Drejer, A Schoesboe, NH Diemer. Regional cerebral glucose

phosphorylation and blood flow after insertion of a microdialysis fiber through the dorsal hippocampus in the rat. J Neurochem 49:729–734, 1987.

24. BHC Westerink, JB De Vries. Characterization of in vivo dopamine release as determined by brain microdialysis after acute and subchronic implantations: Methodological aspects. J Neurosci 51:683–687, 1988.

25. D Allen, PA Crooks, RA Yokel. 4-Trimethylantipyrine: A quaternary ammonium nonradionuclide marker for BBB integrity during in vivo microdialysis. J Pharm Methods 28:129–135, 1992.

26. ECM de Lange, M Danhof, AG de Boer, DD Breimer. Critical factors of intracerebral microdialysis as a technique to determine the pharmacokinetics of drugs in rat brain. Brain Res 666:1–8, 1994.

27. I Westergren, B Nystrom, A Hamberger, BB Johansson. Intracerebral microdialysis and the blood–brain barrier. J Neurochem 64:229–234, 1995.

28. ECM de Lange, M Danhof, AG de Boer, DD Breimer. Methodological considerations of intracerebral microdialysis in pharmacokinetic studies on blood–brain barrier transport of drugs. Brain Res Rev 25:27–49, 1997.

29. O Major, T Shdanova, L Duffek, Z Nagy. Continuous monitoring of blood–brain barrier opening to ^{51}Cr-EDTA by microdialysis following probe injury. Acta Neurochir S51:46–48, 1990.

30. U Ungerstedt. Measurement of neurotransmitter release by intracranial dialysis. In: CA Marsden, ed. Measurement of Neurotransmitter Release In Vivo. New York: John Wiley, 1984, pp 210–245.

31. H Benveniste, PC Huttemeier. Microdialysis, theory and application. Prog Neurobiol 35:195–215, 1990.

32. PM Bungay, PF Morrison, RL Dedrick. Steady-state theory for quantitative microdialysis of solutes and water in vivo and in vitro. Life Sci 46:105–119, 1990.

33. PF Morrison, PM Bungay, JK Hsiao, BA Ball, IN Mefford, RL Dedrick. Quantitative microdialysis: Analysis of transients and application to pharmacokinetics in brain. J Neurochem 57:103–119, 1991.

34. P Lonnröth, PA Jansson, U Smith. A microdialysis method allowing characterization of intercellular water space in humans. Am J Physiol 253: (Endocrinol Metab 16) E228–E231, 1987.

35. RJ Olson, JB Justice. Quantitative microdialysis under transient conditions. Anal Chem 65:1017–1023, 1993.

36. D Scheller, J Kolb. The internal reference technique in microdialysis: A practical approach to monitoring dialysis efficiency and to calculating tissue concentrations from dialysate samples. J Neurosci Methods 40:31–38, 1991.

37. Y Wang, SL Wong, RJ Sawchuk. Comparison of in vitro and in vivo calibration of microdialysis probes using retrodialysis. Curr Sep 10:87, 1991.

38. RA Yokel, DD Allen. Antipyrine as a dialyzable reference to correct differences in efficiency among and within sampling devices during in vivo microdialysis. J Pharm Methods 27:135–142, 1992.

39. SL Wong, K Van Belle, RJ Sawchuk. Distributional transport kinetics of zido-

vudine between plasma and brain extracellular fluid and cerebrospinal fluid blood barriers in the rabbit: Investigation on the inhibitory effect of probenecid utilizing microdialysis. J Pharmacol Exp Ther 265:R1205–R1211, 1993.

40. ECM de Lange, C Zurcher, M Danhof, AG de Boer, DD Breimer. Repeated microdialysis perfusions: Periprobe tissue reactions and BBB permeability. Brain Res 702:261–265, 1995.

41. ECM de Lange, MR Bouw, JW Mandema, M Danhof, AG de Boer, DD Breimer. Application of intracerebral microdialysis to study regional distribution kinetics of drugs in rat brain. Br J Pharmacol 116:2538–2544, 1995.

42. ECM de Lange, MB Hesselink, M Danhof, AG de Boer, DD Breimer. The use of intracerebral microdialysis to determine changes in blood–brain barrier transport characteristics. Pharm Res 12:129–133, 1995.

43. ECM de Lange, JD De Vries, C Zurcher, M Danhof, AG de Boer, DD Breimer. The use of intracerebral microdialysis for the determination of pharmacokinetic profiles of anticancer drugs in tumor-bearing rat brain. Pharm Res 12:1924–1931, 1995.

44. F Thiebaut, T Tsuruo, H Hamada, MM Gottesman, I Pastan, MC Willingham. Immunohistochemical localization in normal tissue of different epitopes in multidrug transport protein P170: Evidence for localization in brain capillaries and crossreactivity of one antibody with a muscle protein. J Histochem Cytochem 37:159–164, 1989.

45. I Sugawara, H Hirofumi, T Tsuruo, S Mori. Specialized localization of P-glycoprotein recognized by MRK-16 monoclonal antibody in endothelial cells of the brain and the spinal cord. Jpn J Cancer Res 81:727–730, 1990.

46. AH Schinkel, E Wagenaar, L Van Deemter, CAAM Mol, P Borst. Absence of the *mdrla* P-glycoprotein in mice affects tissue distribution and pharmacokinetics of dexamethasone, digoxine, and cyclosporin A. J Clin Invest 96:1698–1705, 1995.

47. JS Lee, M Alvarez, C Hose, A Monks, M Grever, AT Fojo, SE Bates. Rhodamine efflux patterns predict P-glycoprotein substrates in the National Cancer Institute drug screen. Mol Pharmacol 46:627–638, 1994.

48. ECM de Lange, G de Bock, AH Schinkel, AG de Boer, DD Breimer. BBB transport and P-glycoprotein functionality using *mdrla* (−/−) and wild type mice. Total versus microdialysis concentration profiles of rhodamine-123. Pharm Res 15(11):1657–1665, 1998.

49. PM Bungay, PF Morrison, LD Dedrick. Steady-state theory for quantitative microdialysis of solutes and water in vivo and in vivo. Life Sci 46:105–119, 1989.

50. A Schnecko, M Pons, K Witte, A Schänzer, J Cambar, B Lemmer. Circadian rhythms in the renin–angiotensin system, renal excretion and hemodynamics in transgenic hypertensive rats. Naunyn-Schiedebergs Arch Pharmacol 69:353R, 1996.

51. B Lemmer, A Matters, M Böhm, D Ganten. Circadian blood pressure variation in transgenic hypertensive rats. Hypertension 22(1):97–101, 1993.

52. ECM de Lange, AG de Boer, DD Breimer. Microdialysis for pharmacokinetic analysis of drug transport to the brain. Adv Drug Delivery Rev 36:211–227, 1999.

53. R Masereeuw, MM Moons, FGM Russel. Rhodamine-123 accumulates extensively in the isolated perfused rat kidney and is secreted by the organic cation system. Eur J Pharmacol 321(3):315–323, 1997.

54. ICJ van der Sandt, AG de Boer, DD Breimer. Unpublished results, 1999.

6

The Role of Brain Extracellular Fluid Production and Efflux Mechanisms in Drug Transport to the Brain

David. J. Begley, Ehsan Ullah Khan, Christopher Rollinson, and Joan Abbott
King's College London, London, England

Anthony Regina and Françoise Roux
INSERM U26, Hôpital Fernand Widal, Paris, France

I. INTRODUCTION

The in vivo penetration of drugs into the central nervous system (CNS) is usually quantified by experimentally determining the brain content of drug with time to derive an influx constant (microliters per milligram of tissue per minute) or permeability coefficient (centimeters per second). Most in vivo experimental methods describing drug uptake into the brain will automatically incorporate any activity of CNS efflux into their apparent determination of brain penetration. Active efflux from the CNS via specific transporters may often reduce the measured penetration of a drug at the blood–brain barrier (BBB) to levels that are lower than might be predicted from the physicochemical properties of the drug, for example, its lipid solubility (1) (Fig. 1).

Within the central nervous system are a number of efflux mechanisms that will influence drug concentrations within the brain. Some of these mechanisms are passive and some active. These efflux mechanisms are central to

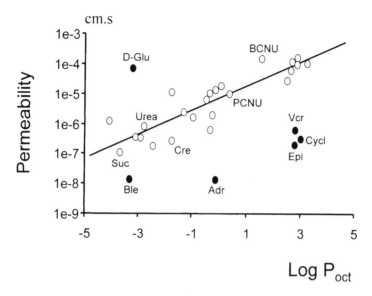

Fig. 1 Plot of CNS permeability against log $P_{octanol}$. Many solutes (open circles) show a clear correlation between their lipid solubility, determined as log P_{oct} and CNS penetration; Suc, sucrose; Cre, creatinine; PCNU, (1-(2-1-nitrosourea; BCNU, 1,3-bis-chloro(2-chloroethyl)1-nitrosourea. Solutes that show an enhanced or depressed uptake at the BBB in relation to their lipid solubility are distinguished as marked outliers on this type of plot (solid circles) and either have a facilitated penetration at the BBB such as D-Glu (D-glucose) or an active efflux from the CNS as in the case of Ble (bleomycin), Adr (Adriamycin), Epi (epipodophyllotoxin/etoposide), Cycl (cyclosporin A), and Vcr (vincristine). (Adapted from Ref. 1.)

the maintenance of homeostasis within the extracellular fluid (ECF) compartments of the brain, the constancy of which is essential for normal neuronal and synaptic function. The activity of these efflux mechanisms also significantly influences the concentration in brain extracellular fluid of free drugs that are available to interact with drug receptor sites. The protein content of cerebrospinal fluid (CSF) and brain extracellular fluid is low compared to plasma, some 20 mg/100 ml (2), and thus protein binding is not a significant factor in maintaining the overall concentration of drugs in these fluids. However, the degree to which a drug may enter cells within the CNS will have an important influence on the free extracellular fluid concentration of drug and the total brain content.

II. ACTIVE SECRETION OF BRAIN EXTRACELLULAR FLUID

A fundamental process that will affect the concentration of all solutes in brain ECF and CSF is the continuous turnover of these compartments as the result of secretion, bulk flow, and drainage of fluid. Extracellular fluid is secreted continuously across the epithelial cells of the choroid plexuses and the endothelial cells of the brain microvasculature into the CSF and extracellular spaces of the brain, respectively. Both cell layers possess tight cell junctions and thus form a barrier to free diffusion. This fluid secretion is primarily the result of Na^+/K^+-ATPase activity at the cell membranes facing the extracellular fluid compartments, which extrude sodium from the cell (3, 4). Thus brain extracellular fluid is constantly being secreted by the brain microvasculature, moving by bulk flow through the interstitium, where it enters the ventricles across the ependyma to combine with CSF or across the pia mater to join the subarachnoid CSF before draining to venous blood through the arachnoid granulations.

Brain extracellular fluid secretion will have important effects on the concentration of solutes that passively cross the blood–brain barrier. If we take substances that are relatively lipid insoluble, thus remaining largely extracellular, and experimentally maintain their plasma levels steady by constant intravenous infusion, the system will tend to a steady state with concentrations in brain interstitial fluid (ISF) and CSF less than those found in plasma. For example, Hollingsworth and Davson (5) infused radiolabeled sulfate, which because of its charge penetrates into the brain relatively slowly, intravenously into rabbits. At all time points the concentration of sulfate is higher in brain extracellular fluid than it is in CSF (Fig. 2), and after 180 minutes of infusion the space occupied by the sulfate is 8% for brain extracellular fluid and 5% in CSF.

If the CNS is considered to be a simple three-compartment model (Fig. 3) consisting of blood, brain ECF, and CSF compartments (6) within which a solute, whose plasma concentration is maintained constant by intravenous infusion, is allowed to diffuse across the semipermeable barriers into brain and CSF, the movement of solute into brain can be followed. Initially the influx into brain will be greater than back-diffusion; as a steady state is approached, however, the fluxes will become equal, and given sufficient time, the plasma concentration, the brain extracellular fluid concentration, and the CSF concentration of solute would all become the same. Substances that penetrate most rapidly into brain would reach this steady state condition in a shorter time. If we now introduce the secretion and bulk flow of fluid in the brain

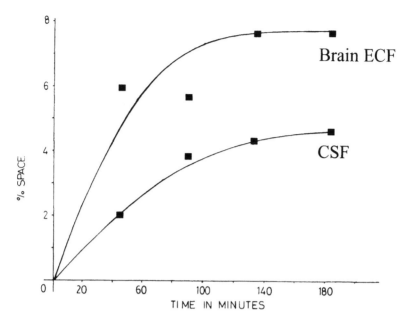

Fig. 2   Penetration of [³⁵S]SO₄ from blood to CNS. Radiolabeled sulfate was infused intravenously into rabbits to maintain a constant plasma concentration. At differing times CSF was withdrawn and brain tissue removed and the [³⁵S]SO₄ concentration in both determined. The brain ECF concentration is corrected for the brain extracellular space (20%). (From Ref. 5.)

ECF and CSF compartments, the steady state equilibrium concentration of solute in these compartments, will never equal that of plasma, even if we allow infinite time. Even substances that enter brain relatively rapidly will be affected to some extent by this phenomenon.

There is a more complex set of reasons to explain why the brain ECF levels of a tracer remain higher than the CSF levels. The CSF compartment turns over more rapidly than the ISF compartment, thus having a greater effect on the reduction in solute concentration in the compartment. Table 1 shows calculated CSF turnover times for several species. If the rate of brain ECF formation is approximately 0.17 µl per gram of brain per minute (7), then a human brain weighing some 1200 g will produce 200 µl of ECF per minute, or 12 ml an hour. For extracellular space of brain tissue of 20%, this would represent approximately 240 ml of ECF for the human brain, giving a turnover

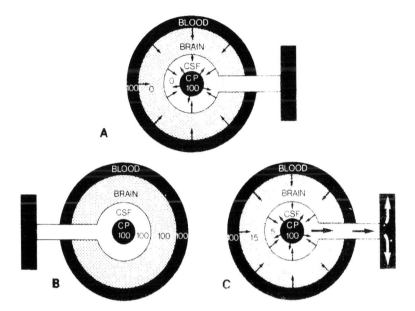

Fig. 3 Sink action of brain ECF and CSF bulk flow. The solid rectangular compart-ments represent blood; the peripheral annuli (solid) represent blood in brain capillaries, and the central (solid) spot, blood in the choroid plexuses. The stippled area represents brain ECF and the open area the CSF. (A) Initial state, where the concentration of the extracellular marker is equal to 100 in blood plasma and 0 in the other compartments. (B) At infinite time, assuming no turnover of brain ECF and CSF. (C) At infinite time with turnover in the brain ECF and CSF compartments. (From Ref. 6.)

time for brain extracellular fluid of about 20 hours, compared to 4.4 hours for ventricular CSF (see Table 1). In addition, the surface area of the BBB microvascular endothelium available for solute exchange is much larger in relation to the ECF volume than the epithelial surface area of the choroid plexus in relation to CSF volume, again contributing to the concentration dif-ferences.

Therefore, for many solutes there is a permanently maintained concen-tration gradient between brain interstitial fluid and CSF. This phenomenon is termed the "sink effect" of the CSF, inasmuch as the CSF constitutes a contin-uous sink for solutes in brain interstitial fluid.

As mentioned previously, a significant proportion of the total production of brain extracellular fluid comes from sources other than the choroid plexuses

Table 1 Rates of Production and Turnover of CSF

Species	Rate of production ($\mu l/min$)[a]	Percent turnover ($\%/min$)[a]	Turnover (h)[b]
Mouse	0.33	0.89	1.87
Rat	2.10	0.72	2.31
Rabbit	10.00	0.43	3.88
Cat	20.00	0.45	3.70
Sheep	118.00	0.83	2.01
Human	350.00	0.38	4.39

[a] From Ref. 2.
[b] Calculated from columns 2 and 3. Assuming a constant rate of brain ECF production per gram of brain between species (the capillary density and capillary surface area per unit weight of brain appear to be relatively constant between species), the ECF production rate and turnover time will be proportional to brain mass and ECF turnover similar in most species (≈ 20 h). The CSF compartment thus turns over faster than the brain ECF compartment.

(7). A continuous production of fluid at the blood–brain barrier itself, which results in the bulk flow of interstitial fluid through the brain interstitium, has the effect of evening out any concentration differences that might occur within the interstitial fluid compartment as a result of diffusive movement. Bulk flow of brain interstitial fluid also means that drugs of larger molecular weight, which may be injected directly into brain substance or released into the CNS from slow-release implants, will be continually carried away from their point of application. The secretion-induced movement of brain ECF will carry drugs through the brain extracellular space at a rate that is largely independent of their molecular weight (8).

The continual turnover of brain interstitial fluid and CSF has important implications for the CNS concentrations of drug that may be achieved following both passive and active penetration of drugs into the brain across the blood–brain barrier. Because of these sink effects, drug concentrations in CSF will usually be lower than those in the brain interstitial fluid, and if a drug is sequestered within neurones and glia, the free drug levels in these fluids may not reflect the overall brain content of drug. Thus CSF sampling or determination of free drug level by intracerebral microdialysis may not reflect the total brain content of an administered drug.

III. ACTIVE DRUG EFFLUX FROM BRAIN

There are a number of specific and active efflux transport mechanisms present in the CNS located both in the endothelial cells forming the blood–brain barrier and in the choroid plexuses. These mechanisms transport endogenous substrates, xenobiotics and also drugs, from the cerebral compartment to blood.

A number of active transport mechanisms have now been described which transport physiologically and pharmacologically active substances out of the brain and into blood. These will have the effect of reducing the concentrations of these compounds in brain and reducing their effective penetration into brain from blood. Recently much attention has been focused on the so-called multidrug transporters; multidrug resistance protein (MRP), P-glycoprotein (Pgp), and the multispecific organic anion transporter (MOAT). All three transporters are members of the ABC cassette (ATP-binding cassette) of transport proteins (9), the members of which show a considerable degree of structural homology. The structural and functional relationships between these transporters and their expression in different species are complex.

P-glycoprotein (Pgp) is the product of the MDR gene in humans and the *mdr* gene in rodents. In humans there are two MDR genes *MDR1* and *MDR2*. The *MDR1* gene product is capable of multidrug transport. In rodents there are three *mdr* genes: *mdr1a, mdr1b*, and *mdr2*. The gene products *mdr1a* and *mdr1b* are capable of multidrug transport. In the rodent *mdr1a* has been shown to be predominantly expressed in the endothelial cells forming the blood–brain barrier (10). The functions of the human *MDR2* and the rodent *mdr2* are less clearly defined and do not appear to be concerned with the phenomenon of multidrug resistance. They may be concerned with phospholipid transport in the liver.

More recently another gene product, the expression of which can confer multidrug resistance, is the multidrug resistance–associated protein (MRP) (11). In humans there appear to be five isoforms of the MRP gene product *MRP1, MRP2, MRP3, MRP4*, and *MRP5*, and there are apparently different levels of expression of these various isoforms in different tissues (12). In rodents only an *mrp1* and an *mrp2* have been distinguished (13). In human and rat liver *MRP2* and *mrp2* the term "canalicular multispecific organic anion transporter" (cMOAT) has been applied (14).

P-glycoprotein accepts a wide range of lipid-soluble substrates and will actively efflux these from cells expressing the gene product. Substrates for *mdr1a* include ivermectin, vinblastine, morphine, dexamethasone, digoxin, cyclosporin A, ondansetron, and loperamide (15–17). *MRP1* accepts a similarly wide variety of substrates but will also tolerate a higher degree of ionization and hydrophilicity in the molecule. Expression of *MRP1* by cells confers a resistance to anthracyclines, vinca alkaloids, and epipodophyllotoxin (etoposide), (18, 19), and these substances would thus appear to be substrates for *MRP1*. It has also been suggested that reduced glutathione (glutathione disulfide) and glutathione conjugates, as well as glucuronidates, may be preferred substrates for *MRP1* (20–22). There does, however, appear to be a substantial overlap in the substrates tolerated by Pgp and MRP.

A multispecific organic acid transporter in the choroid plexus of the fourth ventricle was described a number of years ago by Pappenheimer et al. (23). The organic acid transporter of the choroid plexus, which shows some similarity in its substrate preferences with MRP, may be an isoform or subtype of MRP. Pappenheimer showed that it transported the iodinated X-ray contrast medium Diodrast and phenolsulfonephthalein. This transport could be inhibited by probenecid and *p*-aminohippurate (PAH). MOAT in the choroid plexus is strongly inhibited by probenecid, which also inhibits MRP activity (24). The transport of penicillin from the ventricular CSF is also significantly inhibited by probenecid (25), and thus it seems highly likely that the β-lactam antibiotics are also removed from brain by this mechanism.

Bárány (26) described two MOATs in liver and kidney, the predominant liver transporter showing a strong preference for the substrate iodipamide with a weak inhibition by 2-iodohippuric acid (*o*-iodohippurate) (27, 28). The reverse is the case for the kidney MOAT. The transport mechanism for iodipamide is highly specific for that substrate. The MOAT in the choroid plexus also shows a strong preference for the substrate iodipamide (27, 28), suggesting that choroid plexus MOAT most resembles the liver transport system, the so-called canalicular MOAT (cMOAT), and is thus most likely *MRP2*. Iodipamide and 2-iodohippuric acid were used as X-ray contrast media for many years to assist researchers in visualizing the biliary tract and the kidney, respectively. This strategy was based on the different transport properties of these media, for which the underlying mechanisms are now apparent. Specific functions have not yet been attributed to the other gene products, *MRP3*, *MRP4*, and *MRP5* (12).

Recently Regina et al. (29) made a study of the expression of P-glycoprotein and MRP in the rat. The expression of *MRP1*, *mdr1a*, and *mdr1b* have been investigated by the reverse transcriptase polymerase chain reaction (RT-PCR) and Western blotting in brain homogenate, freshly isolated brain microvessels, primary cerebral endothelial cell cultures, and immortalized rat brain cerebral endothelial cell cultures, RBE4 (30). Functional activity of the

Table 2 Relative Expression of MRP and Pgp

	Homogenate	Isolated microvessels	Primary endothelial cells	Immortalized RBE4
RT-PCR[a]				
MRP mRNA	+	+	+	+
mdr1a mRNA	+	+	+	+
mdr1b mRNA	+	−	+	+
Western blotting[b]				
MRP	+ +	+	+ + +	+
Pgp	±	+ + +	+ +	+
Vincristine accumulation[c]				
MRP activity			+ +	±
Pgp activity			+ +	+

[a] Reveals the presence of mRNA signal for *MRP* and *mdr1a* and *mdr1b*.
[b] Reveals the presence of protein gene product.
[c] A measure of the functional expression of transporter protein.
Source: Ref. 29.

efflux mechanisms has been assessed in vitro in cultured cerebral endothelial cells. The results of this study are summarized in Table 2. The analysis by RT-PCR reveals the presence of messenger RNA for all three transporters in all samples, with the notable exception of *mdr1b* in the freshly isolated microvessels. The presence of the *mdr1b* signal in homogenate and its absence in isolated brain microvessels suggest that *mdr1b* expression may be by cells other than the endothelial cells that form the blood–brain barrier and may be expressed diffusely in both neurones and glia. In both primary endothelial cell cultures and immortalized endothelial cells, the *mdr1b* mRNA signal is present. Western blotting, which gives semiquantitative data, shows a strong *MRP1* signal in brain homogenate but relatively less in isolated microvessels than Pgp, suggesting that *MRP1* activity is largely located in brain regions other than the blood–brain barrier. *MRP1* expression appears to be up-regulated in primary endothelial cell cultures. Complementary functional studies have also been conducted on vincristine accumulation by primary-cultured endothelial cells and in the immortalized rat cerebral endothelial cells (RBE4) (see Table 2). The Pgp substrate cyclosporin A enhanced the accumulation of vincristine in both primary-cultured cells and RBE4 cells, whereas inhibitors of MRP activity, such as genestein, indicate a greater MRP activity in primary cultures, an observation suggesting its up-regulation in primary cell cultures. Interestingly, probenecid also produces a significant increase in vincristine accumulation in primary-cultured cells, suggesting some MOAT-like activity.

The expression and location of Pgp in rat brain tissue has been investigated at the cellular level by means of an electron microscopical technique (31). This study utilizes a monoclonal antibody (C219), directed at an intracellular epitope of all Pgp molecules, and a double-antibody, silver amplification technique to visualize the Pgp. The tissues are permeabilized by prefixation and saponin treatment, affording the antibody access to the intracellular epitope of the Pgp molecule. The technique reveals a strong Pgp localization associated with luminal membrane of the cerebral capillaries; but in the permeabilized tissue, there is no silver reaction present in neurones and glia or glial endfeet in brain substance. Thus even these highly sensitive double-antibody techniques are not powerful enough to reveal the lower expression of *mdr1b* in structures other than capillaries where a signal can be detected with RT-PCR. This observation is compatible with the weak Pgp signal obtained by Western blotting by Regina (29) for cortical homogenate and the very strong signal from isolated cerebral capillaries, derived from the same amount of protein loaded onto the gel. A high power electron microscopical cross section of a cerebral capillary (Fig. 4) clearly shows Pgp in association with the luminal membrane of the capillary endothelial cells, and no silver grains are associated with astrocytic endfeet (31). In 1997 Pardridge et al. (32) suggested that in human tissue samples the predominant expression of Pgp in brain may be localized in the astrocytic endfeet. This suggestion is not supported by the electron microscopical evidence above and the previously described RT-PCR and Western blotting studies of Regina et al. Indeed, to fulfill one of its major functions as a protective mechanism for the brain against lipophilic xenobiotic neurotoxins, it is essential that the predominant Pgp expression be at the luminal membrane of the capillary endothelium, thus presenting a first line of defense for the brain (33).

Khan et al. (34) have recently performed a series of functional studies on the uptake of a number of substances into RBE4 cells in relation to their lipophilicity and reactivity with P-glycoprotein. Pgp activity was assessed by means of a functional method based on the accumulation of [^{3}H]colchicine into RBE4 cells (33). The net accumulation of [^{3}H]colchicine with time by the RBE4 cells is a combination of the passive diffusive influx of colchicine countered by the active efflux of the tracer drug by Pgp. Drugs and compounds introduced into the incubation medium as competitive or noncompetitive inhibitors of Pgp will reduce the efflux component of this process and result in an enhanced intracellular accumulation of [^{3}H]colchicine with time. If 50 µM

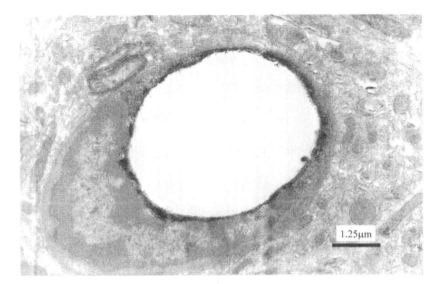

Fig. 4 Electron micrograph showing a cross section of the silver double-antibody amplification (DAB) technique using the monoclonal antibody C219 in permeabilized rat brain. The silver grains indicating the reaction product with Pgp lie immediately below the luminal membrane of the capillary endothelial cells. No reaction product is associated with the abluminal membrane of the endothelial or glial cells or the structures adjacent to the endothelial cells. (From Ref. 31.)

of a number of test drugs is applied to cells grown to confluence in 96-well tissue culture plates, and the uptake of colchicine into the cells expressed as a distribution volume (microliters per milligram of protein), the percentage increase in colchicine accumulation at a defined time point may be directly related to the potency of the drug as an inhibitor of Pgp. [The lipid solubility of the drugs was obtained from the MedChem Database as the log D (octanol) at pH 7.4.]

The results (see Fig. 5) demonstrate that for most of the compounds studied in this set, there is a direct relationship between log D_{oct} and their ability to inhibit Pgp activity (34). Inasmuch as lipid solubility is directly related to the ease with which a drug can intercalate with the cell membrane, this finding is in good agreement with the recent suggestion of Stein (35) that the interaction of Pgp with many of its substrates may take place within the plasma membrane. It would also appear that with the small drug set used in this study, the drugs fall clearly into two groups: those whose lipid solubility

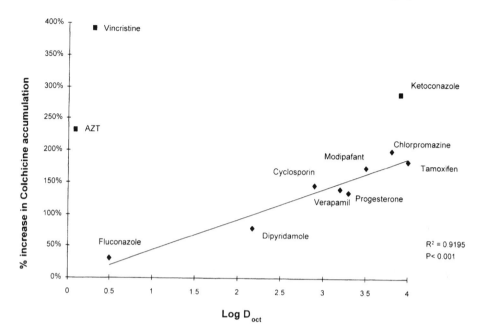

Fig. 5 Relationship of the lipid solubility of a number of drugs (log D_{oct} at pH 7.4) with the percentage increase in the colchicine accumulation of [³H]colchicine by RBE4 cells. The drugs represented by diamonds interact with Pgp in a manner that increases in direct proportion to their lipid solubility. The regression analysis is based on these values. The drugs represented by the squares inhibit Pgp activity to a far greater extent than might be suggested by their lipid solubility, suggesting that they interact with Pgp in a different manner ($n = 6$ for all points). (From Ref. 34.)

is very closely related to their ability to inhibit Pgp, and those like AZT, ketoconazole, and vincristine, which exhibit a significantly greater inhibitory activity than their lipid solubility might suggest. For the drugs that inhibit Pgp in relation to their lipid solubility, the correlation holds good in spite of the remarkable structural diversity of the set of compounds used. Thus for this group of compounds the drug–Pgp interaction may be determined by the concentration of drug available to Pgp within the cell membrane. The more lipid soluble a drug, the longer its dwell time in the plasma membrane is likely to be, and the greater its chances of interacting with Pgp. This emphasis on lipid solubility, rather than a stereochemical one, may well be the reason for the great difficulty of using Pgp to establish structure–affinity relationships for substrates. For the small group of compounds used in this study, the interaction cannot be explained solely on the basis of lipophilicity, and some of these compounds (AZT, ketoconazole, vincristine) may have structural features that are recognized stereochemically by Pgp with a high affinity, hence their appearance as outliers from the line of correlation. An alternative explanation may be that Pgp transports these substances from the cytosolic compartment of the cell rather than from the interior of the lipid membrane employing a different active site on the Pgp molecule.

IV. CONCLUSION

A number of mechanisms in the central nervous system act together in parallel and affect the effective penetration of drugs into brain. These mechanisms will affect the measured brain distributions of a range of compounds and will result in actual brain uptakes lower than might be predicted from physico-chemical data. Strategies directed at increasing brain uptake of drugs that are substrates for specific efflux mechanisms need to be focused on designing reactivity with a transporter out of the drug molecule or by examining ways of inhibiting the activity of an efflux mechanism by coadministering a competitive or noncompetitive inhibitor of the efflux pump together with the desired drug. To be fully and successfully exploited, both strategies will require a greater knowledge of how these efflux mechanisms function and their precise locations within the CNS. A corollary of this line of reasoning is that combination therapies, in which two or more drugs in a mixture are substrates or inhibitors for an active efflux mechanism, may result in an unexpectedly high level of a potentially neurotoxic drug in the CNS as a result of interaction with these transporters.

REFERENCES

1. VA Levin. Relationship of octanol/water partition coefficient and molecular weight to rat brain capillary permeability. J Med Chem 23:682–684, 1980.
2. H Davson, MB Segal. Physiology of the CSF and Blood-Brain Barriers. Boca Raton, FL, CRC Press, 1996, pp 573–574.
3. CE Johanson, SM Sweeney, JT Parmalee, MH Epstein. Cotransport of sodium and chloride by the adult mammalian choroid plexus. Am J Physiol 258:C211–C215, 1990.

4. AL Betz, JA Firth, GW Goldstein. Polarity of the blood–brain barrier: Distribution of enzymes between the luminal and antiluminal membranes of brain capillary endothelial cells. Brain Res 192:17–28, 1980.

5. JG Hollingsworth, H Davson. Transport of sulphate in the rabbit's brain. J Neurobiol 4:389–396, 1973.

6. H Davson. History of the blood–brain barrier concept. In: EA Neuwalt, ed. Implications of the Blood–Brain Barrier. Vol 1. Basic Science Aspects. New York: Plenum Press, 1989.

7. HF Cserr, CS Patlak. Secretion and bulk flow of interstitial fluid. In: MBW Bradbury, ed. Physiology and Pharmacology of the Blood–Brain Barrier. Handbook of Experimental Pharmacology. Vol 103. Berlin: Springer-Verlag, 1992.

8. H Cserr, DN Cooper, PK Suri, CS Patlak. Efflux of radiolabelled polyethylene glycols and albumin from rat brain. Am J Physiol 240:F319–F328, 1981.

9. SPC Cole, G Bhardwaj, JH Gerlach, JE McKemzie, CE Grant, KC Almquist, AJ Stewart, EU Kurz, AMV Duncan, RG Deeley. Overexpression of a transporter gene in a multidrug-resistant human lung cancer cell line. Science 258: 1650–1654, 1992.

10. CR Leveille-Webster, IM Arias. The biology of the P-glycoproteins. J Membr Biol 143:89–102, 1995.

11. T McGrath, MS Center. Adriamycin resistance in HL-60 cells in the absence of detectable P-glycoprotein. Biochem Biophys Res Commun 145:1171–1176, 1987.

12. M Kool, M de Haas, GL Scheffer, RJ Scheper, MJT van Eijk, JA Juijn, F Baas, P Borst. Analysis of expression of cMOAT (*MRP2*), *MRP3*, *MRP4*, and *MRP5*, homologues of the multidrug resistance-associated protein gene (*MRP1*) in human cancer cell lines. Cancer Res 57:3537–3547, 1997.

13. D Keppler, I Leiter, G Jedlitschky, R Mayer, M Buchler. The function of the multidrug resistance proteins (MRP and cMRP) in drug conjugate transport and hepatobiliary excretion. Adv Enzyme Regul 36:17–29, 1996.

14. D Keppler, J Konig, M Buchler. The canalicular multi-drug resistance protein, cMRP/MRP2, a novel conjugate export pump expressed in the apical membrane of hepatocytes. Adv Enzyme Regul 37:321–333, 1997.

15. AH Schinkel, JJM Smit, O van Telligen, JH Beijnen, E Wagenaar, L van Deemter, CAAM Mol, MA van der Valk, EC Robanus-Maandag, HPJ te Tiele, AJM Berns, P Borst. Disruption of the mouse *mdr1a* P-glycoprotein gene leads to a deficiency in the blood–brain barrier and to increased sensitivity to drugs. Cell 77:491–502, 1994.

16. AH Schinkel, E Wagenaar, L van Deemter, P Borst. Absence of the mdr1a P-glycoprotein in mice affects tissue distribution and pharmacokinetics of dexamethasone, digoxin and cyclosporin A. J Clin Invest 96:1698–1705, 1995.

17. AH Schinkel, E Wagenaar, CAAM Mol, L van Deemter. P-glycoprotein in the

blood–brain barrier of mice influences the brain penetration and pharmacological activity of many drugs. J Clin Invest 97:2517–2524, 1996.

18. CE Grant, G Valdimarsson, E Hipfner, KC Almquist, SPC Cole, RG Deeley. Over-expression of multidrug resistance-associated protein (MRP) increases resistance to natural product drugs. Cancer Res 54:357–361, 1990.

19. GJR Zaman, MJ Flens, MR van Leusden, M de Haas, HS Mülder, J Lankelma, HM Pinedo, RJ Scheper, F Baas, HJ Broxterman, P Borst. The human multidrug resistance-associated protein MRP is a plasma membrane drug-efflux pump. Proc Natl Acad Sci USA 91:8822–8826, 1994.

20. I Leiter, G Jedlitschsky, U Buchholz, SPC Cole, RG Deeley, D Keppler. The MRP gene encodes an ATP-dependent export pump for leukotriene C_4 and structurally related conjugates. J Biol Chem 269:27807–27810, 1994.

21. DW Loe, KC Almquist, SPC Cole, RG Deeley. ATP-dependent 17β-estradiol 17-(β-D-glucuronide transport by multidrug resistance protein (MRP). J Biol Chem 271:9683–9689, 1996.

22. G Jedlitschky, I Leiter, U Buchholz, K Barnouin, G Kurtz, D Keppler. Transport of glutathione, glucuronate and sulphate conjugates by the MRP gene-encoded conjugate export pump. Cancer Res 56:988–994, 1996.

23. JR Pappenheimer, SR Heisey, EF Jordan. Active transport of Diodrast and phenolsulfonephthalein from cerebrospinal fluid to blood. Am J Physiol 200:1–10, 1961.

24. R Evers, GJR Zaman, L van Deemter, H Jansen, J Calafat, LCJM Ooman, RPJ Oude Elferink, P Borst, S Schinkel. Basolateral localisation and export activity of the human multidrug resistance-associated protein in polarized pig kidney cells. J Clin Invest 97:1211–1218, 1996.

25. IN Walters, PF Teychenne, LE Claveria, DB Calne. Penicillin transport from cerebrospinal fluid. Neurology 26:1008–1010, 1976.

26. EH Bárány. Inhibition by hippurate and probenecid of in vitro uptake of iodipamide and o-iodohippurate. A composite uptake system for iodipamide in choroid plexus, kidney cortex and anterior uvea of several species. Acta Physiol Scand 86:12–27, 1972.

27. EH Bárány. The liver-like anion transport system in rabbit kidney, uvea and choroid plexus. I. Acta Physiol Scand 88:412–429, 1973.

28. EH Bárány. The liver-like anion transport system in rabbit kidney, uvea and choroid plexus. II. Acta Physiol Scand 88:491–504, 1973.

29. A Regina, A Koman, M Piciotti, B El Hafny, MS Center, R Bergmann, P-O Couraud, F Roux. Mrp1 multidrug resistance–associated protein and P-glycoprotein expression in rat brain microvessel endothelial cells. J Neurochem 71:705–715, 1998.

30. F Roux, O Durieu-Trautmann, N Chaverot, M Claire, P Mailly, J-M Bourre, AD Strosberg, P-O Couraud. Regulation of gamma-glutamyl transpeptidase and alkaline phosphatase activities in immortalised rat brain endothelial cells. J Cell Physiol 159:101–113, 1994.

31. C Rollinson and J Butler. The silver/gold intensification of DAB–chromogen in immunocytochemistry. Microsc Anal March 1997, pp 19–20.

32. WM Pardridge, PM Golden, Y-S Kang, U Bickel. Brain microvascular and astrocyte localization of P-glycoprotein. J Neurochem 68:1278–1285, 1997.

33. DJ Begley, D Lechardeur, Z-D Chen, C Rollinson, M Bardoul, F Roux, D Scherman, NJ Abbott. Functional expression of P-glycoprotein in an immortalised cell line of rat brain endothelial cells. J Neurochem 67:988–995, 1996.

34. EU Khan, A. Reichel, DJ Begley, SJ Roffey, SG Jezequel, NJ Abbott. The effect of drug lipophilicity on P-glycoprotein-mediated colchicine efflux at the blood–brain barrier. Int J Clin Pharm Ther 36:84–86, 1998.

35. WD Stein. Kinetics of the multidrug transporter (P-glycoprotein) and its reversal. Phys Revs 77:545–590, 1997.

7

The Relevance of P-Glycoprotein in Drug Transport to the Brain: The Use of Knockout Mice as a Model System

Ulrich Mayer
University Clinic Eppendorf, Hamburg, Germany

I. INTRODUCTION

P-glycoprotein (Pgp) was first discovered in the 1970s in malignant tumor cells (1). As a cell membrane protein it confers resistance to certain cytostatic agents by removing potentially toxic substances in an energy-dependent transport process as soon as they gain entry to the cells (multidrug resistance, MDR). In the treatment of hematological and oncological malignancies, different cytostatic drugs, including anthracyclines, vinca alkaloids, taxanes, and epipodophyllotoxins, are substrates of the so-called drug-transporting Pgps (MDR1 Pgps) (2, 3). In vitro studies have shown a positive correlation between Pgp expression in malignant tumors and the resistance of the tumors to various chemotherapeutic agents (4–6). In addition to cytostatic substances, many amphiphilic and hydrophobic substances, with widely varying chemical structures, such as the common clinical drugs, digoxin (7), dexamethasone (8), and verapamil (9), are transported by Pgp.

Pgp is expressed not only in malignant tumors but also under physiological conditions in a variety of organs, including kidney (proximal tubular epithelial cells), intestinal mucosa (apical membrane of the epithelial cells), liver (biliary canalicular membrane of the hepatocytes), and brain (apical membrane of brain capillary endothelial cells) (10, 11). From its distribution, the physio-logical function of the MDR1 Pgp in normal organs appears to be the protec-tion of the organism from the toxic action of transported substrates. The devel-opment of Pgp knockout mice has been a major breakthrough contributing to the ongoing elucidation of the function of this protein under in vivo conditions (for review, see Ref. 12).

II. THE Pgp KNOCKOUT MOUSE

In humans there is only a single drug-transporting Pgp (MDR1 Pgp), whereas two Pgps are present in mice: mdr1a (also called mdr3) Pgp and mdr1b (also called mdr1) Pgp (13). Presumably, the two Pgps in mice jointly fulfill the same function as the single MDR1 Pgp in man (14). The distribution of mdr1a Pgp and mdr1b Pgp in mice is organ specific (15), with mdr1a being the exclu-sive Pgp at the blood–brain barrier (BBB), in the intestinal mucosa, and at the blood–testis barrier, whereas mdr1b is the only Pgp detected at the blood–placenta and blood–ovary barriers, and in the cortex of the adrenal glands. Both Pgps are expressed in the kidney and the liver. Three strains of knockout mice [*mdr1a* (−/−) mice, *mdr1b* (−/−) mice, *mdr1a/1b* (−/−) mice] were generated by homozygous disruption of the *mdr1a* or *mdr1b* gene or both. Under laboratory conditions, all three strains of knockout mice showed normal life expectancy, were fertile, and yielded no abnormal histological or labora-tory chemistry findings (16, 17). Direct intestinal Pgp-mediated secretion of transported substrates from systemic circulation into the intestinal lumen was demonstrated in pharmacokinetic investigations (e.g., with digoxin, paclitaxel) (18, 19). In other tissues (adrenals, testes, ovaries), in which Pgp is expressed under physiological conditions, elevated concentrations of Pgp substrates in mdr1a Pgp- or mdr1b Pgp-deficient knockout mice were interpreted as indica-tive of defective organ protection (16, 17, 20).

The following discussion focuses on the function of Pgp in the apical membrane of brain capillary endothelial cells at the BBB with a particular focus on the significance of the loss of functional Pgp for drug transport to the central nervous system (CNS).

III. INVESTIGATION OF THE IN VIVO FUNCTION OF Pgp
AT THE BLOOD–BRAIN BARRIER USING Pgp
KNOCKOUT MICE

The studies on Pgp knockout mice [initially conducted on the *mdr1a* (−/−) mice] originated with the observation that Pgp is an essential component of the intact BBB in vivo (16). In the *mdr1a* $(-/-)$ mice, deficient for Pgp in the apical membrane of brain capillary endothelial cells, the LD_{50} of the anthelmintic agent ivermectin, a known Pgp substrate, was 100-fold lower than in wild-type mice. This finding may be explained by the results of pharmacokinetic studies, in which 100-fold higher concentrations of radiolabeled ivermectin were detected in brain tissue of *mdr1a* $(-/-)$ mice than in wild-type mice. Meanwhile, several other substances have been identified, the levels of which were substantially increased in the brain tissues of knockout mice (compared to wild-type mice) following intravenous administration (see Table 1) (16, 20–27). Such results are of special clinical interest, since the vast majority of these substances are commonly prescribed both in clinical and private practice settings. Whereas most of the investigated drugs showed no direct toxic effects to the CNS in short-term pharmacokinetic studies, aside from ivermectin, domperidone (a dopamine antagonist) and loperamide (a common antidiarrheal drug), increased CNS toxicity in short-term studies was observed in knockout mice that were deficient for the mdr1a Pgp (21). Under physiological conditions, as in wild-type mice, no CNS side effects were observed after application of the peripheral opioid loperamide. When loperamide was administered to *mdr1a* $(-/-)$ mice, however, characteristic morphinelike CNS symptoms were observed: hyperactivity, compulsive circling movements followed by sudden immobility, and a typically erect tail on an arched back

Table 1 A Selection of Drugs with an Increased Brain Penetration After Intravenous Injection in Mice Deficient for the mdr1a P-Glycoprotein (*mdr1a* $(-/-)$ or *mdr1a/1b* $(-/-)$ mice)[a]

Drug	Ref.	Drug	Ref.
Digoxin	20	SDZ PSC833	26
Saquinavir	25	Ivermectin	16
Dexamethasone	20, 24	Loperamid	21
Itraconazole	23	Cyclosporin A	21
Vinblastine	22	Domperidone	21
Ondansetron	21	Quinidine	27

[a] Drug concentrations in brain were measured by levels of radioactivity (use of radiolabeled drugs) or determination of the parent compound by specific methods (vinblastine, saquinavir, SDZ PSC833, itraconazole). For domperidone the appearance of CNS side effects in *mdr1a* $(-/-)$ mice was used as an indicator for increased drug penetration into the brain.

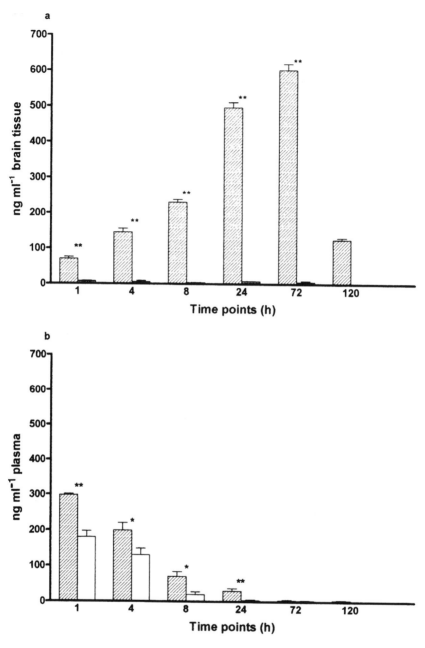

Fig. 1 (a) Brain and (b) plasma levels of radioactivity in *mdr1a* (−/−) mice (hatched columns) and wild-type mice (open columns) after intravenous administration of [³H]digoxin (0.2 mg kg⁻¹). Mean levels with standard error of the mean are depicted (*n* = 3). Brain and plasma levels at 120 hours were not determined for wild-type mice. For each pair of columns, *$p < 0.05$; **$p < 0.01$. (From Ref. 18.)

("straub tail"). Following the administration of radiolabeled loperamide for pharmacokinetic measurements, increased levels of radioactivity were found in the brain of *mdr1a* ($-/-$) mice. The CNS side effects were found to be transient, and the knockout mice fully recovered. In contrast to the short retention time of loperamide in brain tissue, estimated to be on the order of minutes to several hours, the cardiac glycoside digoxin accumulated in the brain during long-term studies (18). Within the first 72 hours postadministration of an intravenous bolus of radiolabeled digoxin, the levels of radioactivity in the brain of *mdr1a* ($-/-$) mice increased by up to 200-fold over the levels detected in wild-type mice at a point in time when the plasma concentrations had already returned to levels close to the detection limit (Fig. 1). A similar accumulative effect was observed with the cytostatic agent vinblastine (22). Neither substance showed any evidence of CNS-toxic effects in the knockout mice. However, especially with the vinca alkaloid vinblastine, there are no long-term toxicity studies available to provide results of the examination of histological alterations in the brain.

IV. Pgp REVERSAL AGENTS: USE AND EFFECTS ON THE PENETRATION OF Pgp-TRANSPORTED SUBSTANCES ACROSS THE BLOOD–BRAIN BARRIER

Observations of increased penetration of Pgp substrates across the BBB in knockout mice are of clinical significance, since Pgp inhibitors (so-called reversal agents) have been tested in the past few years in attempts to overcome the multidrug resistance of malignant tumors in man (28–30). Because there are no structural differences between Pgp in tumor cells and Pgp expressed under physiological conditions, any effective reversal agent is bound to also inhibit the natural function of Pgp (31, 32). When Pgp knockout mice are used as a model for the in vivo situation after systemic administration of a reversal agent, substantial impairment of BBB integrity should be expected. Taking into consideration the known, but usually moderate, CNS side effects of digoxin and neurotoxicity of vinblastine in humans, the loss of functional Pgp in the BBB resulting from the application of effective Pgp inhibitors may have severe consequences in the patient. Similarly, the relatively safe application of loperamide in the treatment of diarrhea may be complicated when CNS side effects occur. Indeed, clinical experience exists with fatal toxicity of certain Pgp substrates (e.g., anthracyclines, vinca alkaloids) as a result of increased penetration into brain tissues, although this was a case of accidental intrathecal administration (33, 34). A case report of an organ transplant patient described serious neurological side effects (35) following combined treatment of doxorubicin and chronic administration of cyclosporin A, known to be a very potent Pgp inhibitor (36).

Apart from considering the attenuation of the blood–brain barrier to be an adverse effect of the administration of reversal agents, it must be noted that this event, which also offers the opportunity to transport drugs to the brain directly and unimpeded by BBB-bound Pgp, may thus provide a new approach to the pharmacological therapy of primary CNS disorders. Oncologists may focus on the opportunity of improving the chemotherapeutic treatment of primary brain tumors. To date, malignant CNS tumors not suitable for radical surgery are of poor prognosis, since additional systemic cytostatic therapy has shown little or no effect. In a series of studies on tissue samples from human primary brain tumors, Pgp was shown to be expressed at substantial levels both in newly formed vascular endothelial cells near the location of the tumor and in the tumor cells themselves (37–40). This may result in insufficient transport of the administered cytostatic drugs from the blood circulation to the tumor tissue.

A series of reversal agents is available, many of which are Pgp substrates and show various degrees of efficacy in the inhibition of transport function (41). The calcium antagonist verapamil was the first substance described to be capable of reversing the MDR phenotype of a tumor cell line (42). The in vitro activity of various Pgp reversal agents after coadministration with a cytostatic agent transported by Pgp was rather high. However, the results of the corresponding clinical studies in tumor patients in terms of overcoming multidrug resistance were disappointing (29, 30, 43) and have been the reason for continued development of novel Pgp inhibitors. After the immunosuppressant cyclosporin A, further development and testing of reversal agents has produced substantially more active substances (e.g., the cyclosporin analogue SDZ PSC833, and the carboxin derivative GF120918) (44, 45). In vitro studies have shown the Pgp inhibitory activity of SDZ PSC833 to exceed that of cyclosporin A by a factor of 10 (investigated with Pgp substrates, such as doxorubicin and vincristine) (44). Because of the strong inhibitory activity of these new reversal agents, the in vivo effect on physiologically expressed Pgp is at least as important as the effect on the MDR phenotype of Pgp-expressing tumors.

V. IN VITRO TESTING OF THE INHIBITORY ACTIVITY OF Pgp REVERSAL AGENTS IN KNOCKOUT MICE

The inhibitory activity of SDZ PSC833 and GF120918 on BBB-bound Pgp of wild-type mice was investigated at the Netherlands Cancer Institute (47,

48). The pharmacokinetic properties of radiolabeled digoxin in *mdr1a/1b* (−/−) mice were used as a measure for optimal and selective in vivo inhibition of Pgp. Although significantly elevated digoxin levels were detected in brains of wild-type mice, premedication with one of these two reversal agents failed to establish similar elevated concentrations shown in *mdr1a/1b* (−/−) control mice, not subjected to pretreatment (Table 2). One reason for this apparent incomplete blockade of Pgp in the BBB despite the presence of high plasma levels of the inhibitor might be the extent of plasma protein binding, which is known to be high for many reversal agents (48, 49). However, the reversal agents showed an interesting effect on the pharmacokinetics of Pgp substrates in the *mdr1a/1b* (−/−) mice. After SDZ PSC833 pretreatment, the digoxin brain levels were lower than in *mdr1a/1b* (−/−) control mice. Hence, some combination of extensive plasma protein binding and competition between SDZ PSC833 and the Pgp substrate for potential brain binding sites may be responsible for the observed lack of increase of digoxin levels in the CNS of the wild-type mice. Similar results were obtained when GF120918 was used as the reversal agent (47). Interestingly, in a recent study (50) determining the tissue distribution of radiolabeled verapamil by positron emission tomography, cyclosporin A was shown to completely inhibit Pgp at the BBB in wild-type mice. An additive effect of cyclosporin A on the brain distribution of verapamil in *mdr1a* (−/−) mice was not detected.

Table 2   Plasma and Brain Levels of Radioactivity 4 Hours After Intravenous Administration of [³H]digoxin to Mice Pretreated with SDZ PSC833 or Vehicle Alone

	Levels of radioactivity[a]		
	Plasma (ng/ml)	Brain (ng/g)	Ratio of brain to plasma
Wild-type	11.7 ± 5.0	0.7 ± 0.4	0.060 ± 0.043
Wild-type + SDZ PSC833	28.4 ± 8.1	13.3 ± 5.9	0.47 ± 0.25*
mdr1a/1b (−/−)	28.6 ± 1.4	47.8 ± 7.8	1.67 ± 0.28**
mdr1a/1b (−/−) + SDZ PSC833	35.3 ± 4.6	26.4 ± 1.4	0.75 ± 0.11**

[a] Results are means ± SD ($n = 4$) in nanograms of [³H]digoxin equivalent per milliliter or gram. Wild-type or *mdr1a/1b* (−/−) mice received oral SDZ PSC833 (50 mg/kg) or vehicle alone 2 hours before intravenous injection of [³H]digoxin (0.05 mg/kg). Statistical significance of difference from vehicle treated wild-type mice given in notes * and **.
* $p < 0.02$.
** $p < 0.001$.
Source: Ref. 47.

VI. CONCLUSIONS

The Pgp knockout mice, deficient in mdr1a Pgp at the blood–brain barrier [in *mdr1a* (−/−) and *mdr1a/1b* (−/−) mice], provide an important in vivo model for the elucidation of the function of Pgp in the intact BBB. Seemingly unobstructed penetration of Pgp substrates across the BBB observed in several studies in *mdr1a* (−/−) and *mdr1a/1b* (−/−) mice, shown in several studies for different substrates, has immediate clinical impact. Since increasingly effective Pgp inhibitors are administered to patients in attempts to overcome the resistance of malignant tumors to chemotherapeutic agents, the effects of these reversal agents on physiologically expressed Pgp are gaining in significance. The nonphysiological disruption of the blood–brain barrier effected by inhibition of Pgp with reversal agents may lead to an incalculable CNS toxicity risk when Pgp substrates (e.g., cytostatic agents) are coadministered, but also opens up new routes of pharmacological treatment approaches for primary CNS disorders. The Pgp knockout mice may provide valuable information on the inhibitory activity of a reversal agent at the BBB as well as the substance's Pgp-independent effects, such as competition with a coadministered Pgp substrate for specific binding sites. This may permit detrimental effects of the reversal agent (e.g., when used to overcome MDR in malignant tumors) to be detected prior to the first clinical application of the substance. Moreover, if used with caution, this system may allow determination of the probability of CNS side effects (positive and negative) of Pgp substrates and BBB penetration of these substances after the Pgp function has been blocked.

Finally, it should be noted that the absence of Pgp in the intestinal mucosa of the Pgp knockout mice strongly reduced the excretion of transported substances. In a series of animal experiments and clinical studies, decreased elimination or increased intestinal absorption of Pgp substrates was shown to occur in response to the administration of reversal agents (19, 51, 52). The concomitant increase in area under the plasma curve of the Pgp substrate can also be expected to have a significant influence on the concentration of the substance in various organ systems (e.g., brain).

REFERENCES

1. RL Juliano, V Ling. A surface glycoprotein modulating drug permeability in Chinese hamster ovary cell mutants. Biochim Biophys Acta 455:152–162, 1976.
2. JA Edicott, V Ling. The biochemistry of P-glycoprotein-mediated multidrug resistance. Annu Rev Biochem 58:137–171, 1989.

3. MM Gottesman, I Pastan. Biochemistry of multidrug-resistance mediated by multidrug transporter. Annu Rev Biochem 62:385–427, 1993.
4. J Bourhis, J Bernard, O Hartmann, L Boccon-Gibod, J Lemerle, G Riou. Correlation of MDR1 gene expression with chemotherapy in neuroblastoma. J Natl Cancer Inst 81:1401–1405, 1989.
5. N Baldini, K Scotlandi, G Barbanti-Bródano, MC Manara, D Maurici, G Bacci, F Bertoni, P Picci, S Sottili, M Campanacci, M Serra. Expression of P-glycoprotein in high-grade osteosarcomas in relation to clinical outcome. N Engl J Med 333:1380–1385, 1995.
6. JP Marie, R Zittoun, BI Sikic. Multidrug resistance (MDR1) gene expression in adult acute leukemias: Correlations with treatment outcome and in vitro drug sensitivity. Blood 78:586–592, 1991.
7. Y Tanigawara, N Okamura, M Hirai, M Yasuhara, K Ueda, N Kioka, T Komano, R Hori. Transport of digoxin by human P-glycoprotein expressed in a porcine kidney epithelial cell line (LLC-PK$_1$). J Pharm Exp Ther 263:840–845, 1992.
8. K Ueda, N Okamura, M Hirai, Y Tanigawara, T Saeki, N Kioka, T Komano, R Hori. Human P-glycoprotein transports cortisol, aldosterone, and dexamethasone, but not progesterone. J Biol Chem 267:24248–24252, 1992.
9. M Horio, KV Chin, SJ Currier, S Goldenberg, C Williams, I Pastan, MM Gottesman, J Handler. Transepithelial transport of drugs by the multidrug transporter in cultured Madin–Darby canine kidney cell epithelia. J Biol Chem 264:14880–14884, 1989.
10. F Thiebaut, T Tsuruo, H Hamada, MM Gottesman, I Pastan, MC Willingham. Cellular localization of the multidrug resistance gene product in normal human tissues. Proc Natl Acad Sci USA 84:7735–7738, 1987.
11. C Cordon-Cardo, JP O'Brien, D Casals, L Rittman-Grauer, JL Biedler, MR Melamed, JR Bertino. Multidrug-resistance gene (P-glycoprotein) is expressed by endothelial cells at blood–brain barrier sites. Proc Natl Acad Sci USA 86:695–698, 1989.
12. AH Schinkel. The physiological function of drug-transporting P-glycoproteins. Semin Cancer Biol 8:161–170, 1997.
13. A Devault, P Gros. Two members of the mouse *mdr* gene family confer multidrug resistance with overlapping but distinct drug specificities. Mol Cell Biol 10:1652–1663, 1990.
14. P Borst, AH Schinkel. What have we learnt thus far from mice with disrupted P-glycoprotein genes? Eur J Cancer 32A:985–990, 1996.
15. P Borst, AH Schinkel, JJM Smit, E Wagenaar, L van Deemter, AJ Smith, EWHM Eijdems, F Baas, GJR Zaman. Classical and novel forms of multidrug resistance and the physiological function of P-glycoproteins in mammals. Pharmacol Ther 60:289–299, 1993.
16. AH Schinkel, JJM Smit, O van Tellingen, JH Beijnen, E Wagenaar, L van Deemter, CAAM Mol, MA van der Valk, EC Robanus-Maandag, HPJ te Riele, AJM Berns, P Borst. Disruption of the mouse mdr1a P-glycoprotein gene leads

to a deficiency in the blood–brain barrier and to increased sensitivity to drugs. Cell 77:491–502, 1994.

17. AH Schinkel, U Mayer, E Wagenaar, CAAM Mol, L van Deemter, JJM Smit, MA van der Valk, AC Voordouw, H Spits, O van Tellingen, JJMS Zijlmans, WE Fibbe, P Borst. Normal viability and altered pharmacokinetics in mice lacking mdr1-type (drug-transporting) P-glycoproteins. Proc Natl Acad Sci USA 94: 4028–4033, 1997.

18. U Mayer, E Wagenaar, JH Beijnen, JW Smit, DKF Meijer, J van Asperen, P Borst, AH Schinkel. Substantial excretion of digoxin via the intestinal mucosa and prevention of long-term digoxin accumulation in the brain by the mdr1a P-glycoprotein. Br J Pharmacol 119:1038–1044, 1996.

19. A Sparreboom, J van Asperen, U Mayer, AH Schinkel, JW Smit, DKF Meijer, P Borst, WJ Nooijen, J Beijnen, O van Tellingen. Limited oral bio-availability and active epithelial excretion of paclitaxel (taxol) caused by P-glycoprotein in the intestine. Proc Natl Acad Sci USA 94:2031–2035, 1997.

20. AH Schinkel, E Wagenaar, L van Deemter, CAAM Mol, P Borst. Absence of the mdr1a P-glycoprotein in mice affects tissue distribution and pharmacokinetics of dexamethasone, digoxin, and cyclosporin A. J Clin Invest 96:1698–1705, 1995.

21. AH Schinkel, E Wagenaar, L van Deemter, CAAM Mol. P-glycoprotein in the blood–brain barrier of mice influences the brain penetration and pharmacological activity of many drugs. J Clin Invest 97:2517–2524, 1996.

22. J van Asperen, AH Schinkel, JH Beijnen, WJ Nooijen, P Borst, O van Tellingen. Altered pharmacokinetics of vinblastine in mdr1a P-glycoprotein-deficient mice. J Natl Cancer Inst 88:994–999, 1996.

23. T Miyama, H Takanaga, H Matsuo, K Yamano, K Yamamoto, T Iga, M Naito, T Tsuruo, H Ishizuka, Y Kawahara, Y Sawada. P-glycoprotein-mediated transport of itraconazole across the blood–brain barrier. Antimicrob Agents Chemother 42:1738–1744, 1998.

24. OC Meijer, EC de Lange, DD Breimer, AG de Boer, JO Workel, ER de Kloet. Penetration of dexamethasone into brain glucocorticoid targets is enhanced in mdr1a P-glycoprotein knockout mice. Endocrinology 139:1789–1793, 1998.

25. RB Kim, MF Fromm, C Wandel, B Leake, AJ Wood, DM Roden, GR Wilkinson. The drug transporter P-glycoprotein limits oral absorption and brain entry of HIV-1 protease inhibitors. J Clin Invest 101:289–294, 1998.

26. S Desrayaud, ECM de Lange, M Lemaire, A Bruelisauer, AG de Boer, DD Breimer. Effect of the mdr1a P-glycoprotein gene disruption on the tissue distribution of SDZ PSC 833, a multidrug resistance–reversing agent, in mice. J Pharmacol Exp Ther 285:438–443, 1998.

27. H Kusuhara, H Suzuki, T Terasaki, A Kakee, M Lemaire, Y Sugiyama. P-glycoprotein mediates the efflux of quinidine across the blood–brain barrier. J Pharmacol Exp Ther 283:574–580, 1997.

28. JM Ford, WN Hait. Pharmacology of drugs that alter multidrug resistance in cancer. Pharmacol Rev 42:155–199, 1990.

29. GA Fisher, BI Sikic. Clinical studies with modulators of multidrug resistance. Hematol Oncol Clin North Am 9:363–382, 1995.

30. BI Sikic. Modulation of drug resistance: At the threshold. J Clin Oncol 11:1629–1635, 1993.

31. BL Lum, MP Gosland. MDR expression in normal tissue. Pharmacological implications for the clinical use of P-glycoprotein inhibitors. Hematol Oncol Clin North Am 9:319–336, 1995.

32. BI Sikic, GA Fisher, BL Lum, J Halsey, L Beketic-Oreskovic, G Chen. Modulation and prevention of multidrug resistance by inhibitors of P-glycoprotein. Cancer Chemother Pharmacol 40(suppl.):S13–S19, 1997.

33. PG Bain, PL Lantos, V Djurovic, I West. Intrathecal vincristine: A fatal chemotherapeutic error with devastating central nervous system effects. J Neurol 238: 230–234, 1991.

34. ME Mortensen, AJ Cecalupo, WD Lo, MJ Egorin, R Batley. Inadvertent intrathecal injection of daunorubicin with fatal outcome. Med Pediatr Oncol 20:249–253, 1992.

35. T Barbui, A Rambaldi, L Parenzan, M Zuchelli, N Perico, G Remuzzi. Neurological symptoms and coma associated with doxorubicin administration during chronic cyclosporin therapy. Lancet 339:1421, 1992.

36. JM Ford. Modulators of multidrug resistance. Hematol Oncol Clin North Am 9: 337–361, 1995.

37. LG Feun, N Savaraj, HJ Landy. Drug resistance in brain tumors. J Neurooncol 20:165–176, 1994.

38. K Dietzmann, PV Bossanyi, DS Franke. Expression of P-glycoprotein as a multidrug resistance gene product in human reactive astrozytes and astrozytoma. Zentralbl Pathol 140:149–153, 1994.

39. PM Chou, M Reyes-Mugica, N Barquin, T Yasuda, X Tan, T Tomita. Multidrug resistance gene expression in childhood medulloblastoma: Correlation with clinical outcome and DNA ploidy in 29 patients. Pediatr Neurosurg 23:283–291, 1995.

40. K Toth, MM Vaughan, NS Peress, HK Slocum, YM Rustum. MDR1 P-glycoprotein is expressed by endothelial cells of newly formed capillaries in human gliomas but is not expressed in the neovasculature of other primary tumors. Am J Pathol 149:853–858, 1996.

41. PW Wigler, FK Patterson. Inhibition of the multidrug resistance efflux pump. Biochim Biophys Acta 1154:173–181, 1993.

42. T Tsuruo, H Iida, S Tsukagoshi, Y Sakurai. Overcoming of vincristine resistance in P388 leukemia in vivo and in vitro through enhanced cytotoxicity of vincristine and vinblastine by verapamil. Cancer Res 41:1967–1972, 1981.

43. M Raderer, W Scheithauer. Clinical trials of agents that reverse multidrug resistance. A literature review. Cancer 72:3553–3563, 1993.

44. PR Twentyman, NM Bleehen. Resistance modification by PSC-833, a novel non-immunosuppressive cyclosporin. Eur J Cancer 27:1639–1642, 1991.

45. F Hyafil, C Vergely, P Du Vignaud, T Grand-Perret. In vitro and in vivo reversal

of multidrug resistance by GF120918, an acridonecarboxamide derivative. Cancer Res 53:4595–4602, 1993.

46. U Mayer, E Wagenaar, B Dorobek, JH Beijnen, P Borst, AH Schinkel. Full blockade of intestinal P-glycoprotein and extensive inhibition of blood–brain barrier P-glycoprotein by oral treatment of mice with PSC833. J Clin Invest 100: 2430–2436, 1997.

47. U Mayer, KR Brouwer, JL Jarrett, RC Jewell, EM Paul, JHM Schellens, P Borst, AH Schinkel. The inhibitory activity and selectivity of the reversal agent GF120918 for physiologically expressed P-glycoprotein, using *mdr1a/1b* ($-/-$) mice as an in vivo model for optimal P-glycoprotein inhibition. Fourth International Symposium on Cytostatic Drug Resistance, Berlin, 1997.

48. M Lehnert, R de Gulli, K Kunke, S Emerson, WS Dalton, SE Salmon. Serum can inhibit reversal of multidrug resistance by chemosensitisers. Eur J Cancer 32A:862–867, 1996.

49. AJ Smith, U Mayer, AH Schinkel, P Borst. Availability of PSC833, a substrate and inhibitor of P-glycoproteins, in various concentrations of serum. J Natl Cancer Inst 90:1161–1166, 1998.

50. NH Hendrikse, AH Schinkel, EG de Vries, E Fluks, WT Van der Graaf, AT Willemsen, W Vaalburg, EJ Franssen. Complete in vivo reversal of P-glycoprotein pump function in the blood–brain barrier visualized with positron emission tomography. Br J Pharmacol 124:1413–1418, 1998.

51. BI Sikic, GA Fisher, BL Lum, J Halsey, L Beketic-Oreskovic, G Chen. Modulation and prevention of multidrug resistance by inhibitors of P-glycoprotein. Cancer Chemother Pharmacol 40(suppl.):S13–S19, 1997.

52. JM Meerum Terwogt, JH Beijnen, WW ten Bokkel Huinink, H Rosing, JH Schellens. Co-administration of cyclosporin enables oral therapy with paclitaxel. Lancet 352:285, 1998.

8

Specific Mechanisms for Transporting Drugs Into Brain

Akira Tsuji
Kanazawa University, Kanazawa, Ishikawa, Japan

I. INTRODUCTION

In the development of new drugs, controlling entry into the brain remains a difficult problem. The entry of compounds from the circulating blood into the brain is strictly regulated by the blood–brain barrier (BBB), which is composed of brain capillary endothelial cells (BCECs). Permeation of compounds across the BBB has long been believed to be dependent on their lipophilicity, because the BCECs are linked by tight junctions without fenestrations and form a lipoidal membrane barrier. However, although vincristine, vinblastine, doxorubicin, epipodophyllotoxin, and cyclosporin A are highly lipophilic, the apparent permeation of these drugs across the BBB was unexpectedly low (1–4). This relationship indicates that lipophilicity is not necessarily a useful predictor of the feasibility of transfer into the brain. Furthermore, most nutrients in the circulating blood, in spite of having low lipophilicity, are well known to be efficiently taken up into the brain. These apparently contradictory observations can be ascribed to the existence of multiple mechanisms of drug transport through the BBB (i.e., carrier-mediated influx and/or efflux,) as well as to passive diffusion.

Recent advances in studies on the BBB transport of nutrients, neuroactive agents, and xenobiotics have led to a change in our concept of the BBB. It is no longer regarded as a static lipoidal membrane barrier of endothelial cells, but rather is considered to be a dynamic interface that has physiological

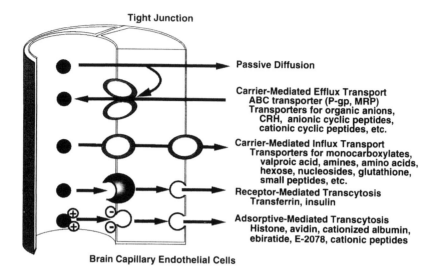

Fig. 1 Blood–brain barrier transport mechanisms.

functions for the specific and selective transmembrane transport of many compounds, as well as degradative enzyme activities. As illustrated in Fig. 1, the BBB has carrier-mediated transport mechanisms working for influx or efflux of endogenous and exogenous compounds, a receptor-mediated transcytosis mechanism specific to certain peptides such as transferrin and insulin, and an adsorptive- (or absorptive-) mediated transcytosis mechanism for positively charged peptides such as histone and cationized albumin (3–7). This chapter provides an overview of transporter-mediated or specialized mechanism–mediated influx and efflux processes at the BBB, to give a basis for understanding the specific net transport of drugs, including large peptides, into the brain. It is not intended to be a review of the literature on drug delivery into the brain by utilizing simple diffusion (e.g., by modifying the lipophilicity of the drugs).

II. ACTIVE EFFLUX TRANSPORT BY P-GLYCOPROTEIN AT THE BLOOD–BRAIN BARRIER

Levin (2) and Pardridge et al. (3) proposed the existence of an upper limit of molecular weight for BBB transport; that is, compounds such as vincristine, doxorubicin, and cyclosporin A, having molecular weight exceeding 400–600

Da, even if they have high lipophilicity, may not be transported across the BBB in pharmacologically significant amounts. In contrast to the hypothesis above (2, 3), it has recently been proposed that active efflux of drugs by P-glycoprotein (Pgp) expressed at the luminal membrane of the brain capillary endothelial cells accounts for the apparently poor BBB permeability to certain drugs (7–17). Results supporting the in vivo importance of Pgp-mediated efflux of several drugs have been obtained in studies using ischemic rat brain (13, 14), brain microdialysis (18, 19) and *mdr1a* gene knockout mice (9, 20–28). As shown in Fig. 2, the concentrations of digoxin, doxorubicin, and cyclosporin A in the brain of *mdr1a* gene knockout mice are significantly higher than those in normal mice, clearly indicating that Pgp at the BBB mediates efflux of these three drugs from the brain (unpublished observations).

The physiological function of Pgp at the BBB for endogenous substrates is of great interest. Pgp-deficient *mdr1a* $(-/-)$ mice have significantly higher brain bilirubin levels than normal mice after intravenous administration of bilirubin (29). These data suggest that bilirubin is a substrate for Pgp and that the increased brain bilirubin content in *mdr1a* $(-/-)$ mice is a result of net brain bilirubin influx due to the decrease of efflux. It is suspected that Pgp expressed at the BBB provides a protective effect against bilirubin neurotoxicity by reducing the net brain bilirubin influx (29). Thus, Pgp at the BBB is thought to be important for limiting access of endogenous and exogenous toxic agents to the brain.

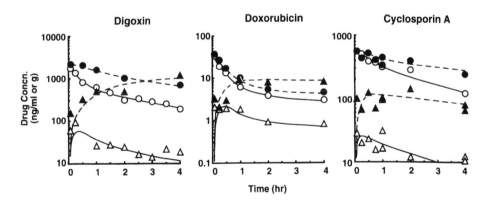

Fig. 2 Time courses of brain distribution of digoxin, doxorubicin, and cyclosporin A after 1 mg/kg intravenous administration in normal (open symbols, solid lines) and *mdr1a* gene knockout mice (solid symbols, dotted line): circles, plasma (or blood for cyclosporin A); triangles, brain. (From Tsuji, unpublished observations.)

Malignant brain tumors often present a difficult therapeutic challenge because anticancer drugs, owing to their limited penetration across the BBB as a consequence of active efflux via Pgp, do not reach therapeutic concentrations in the brain tumor. Increased delivery of anticancer drugs to tumors could improve the survival of patients with malignant brain tumors. In rats inoculated with 9L-glioma cells into the brain, the endothelial cells of tumor-associated vessels ultrastructurally showed increased fenestrations, swelling, and disrupted junctions (30, 31). Therefore, the tumor-associated vessels may be more permeable than the normal vessels in the brain. This expectation was confirmed by the experimental determination of brain uptake index (BUI) values for anticancer drugs in our laboratory according to the method of Oldendorf (32). The BUI value in the 9L-glioma-inoculated region was almost 50% for [^{3}H]fluorouracil, and 28–30% for [^{14}C]ranimustine and doxorubicin. The BUI values of the latter two compounds were essentially the same as the value of 30% for [^{14}C] sucrose. The higher extraction of water-soluble [^{3}H]fluorouracil than that of [^{14}C]sucrose was suggested to be due to carrier-mediated uptake by 9L-glioma cells. In spite of this increased permeability in the tumor region, the BUI values of [^{3}H]fluorouracil, [^{14}C]ranimustine, and doxorubicin were only 2–6%; this is very similar to that of [^{14}C]sucrose in the normal brain region, indicating that the BBB with tight junctions still represents a formidable barrier to anticancer drugs (30).

Because of the active efflux of anticancer drugs by Pgp expressed at the luminal membrane of the BCECs (7–28), the chemotherapy of brain tumors is not very effective. The idea that so-called pharmacological opening of the BBB by inhibition of Pgp function may lead to increased uptake of some drugs including anticancer drugs into brain has been tested in animals. Wang et al. (18) found that the concentration of the P-glycoprotein substrate rhodamine-123 in the brain extracellular fluid, measured by microdialysis, was increased three- to four-fold by intravenous infusion of cyclosporin A. Preintravenous administration of the efficient P-glycoprotein blocker SDZ PSC833, a new nonimmunosuppressive analogue of cyclosporin A, resulted in a five-fold increase of the brain-to-blood concentration ratio of cyclosporin A in rats, whereas the latter compound does not modify the concentration ratio of SDZ PSC833 (33). In an in situ brain perfusion model, the brain uptake of colchicine and vinblastine was increased eight- and nine-fold, respectively, after an intravenous administration of SDZ PSC833 (34). In a continuous intravenous infusion model, SDZ PSC833 enhanced the brain concentration of colchicine at least 10-fold as measured by brain microdialysis, whereas the plasma level was increased only about two-fold (35). Although SDZ PSC833 was at first thought to be not transported by Pgp, a recent study has indicated that both SDS PSC833 and cyclosporin A are substrates of Pgp at the BBB (33).

Extrapolation of the results of these studies indicated that, at a proper dosage, administration of Pgp-blocking agents such as cyclosporin A or SDZ PSC833 to patients with brain tumors may help to enhance the response of their tumors to anticancer drugs such as vinblastine.

III. CARRIER-MEDIATED BLOOD–BRAIN BARRIER TRANSPORT OF DRUGS

There are several transport systems at the BBB for nutrients and endogenous compounds, as described below. Utilization of differences in affinity and maximal transport activity among these transport systems expressed at the BBB is expected to provide a basis for enhancing the delivery or retarding the accumulation of drugs in the brain, as required.

A. Monocarboxylic Acid Transport Systems

The monocarboxylic acid transporters at the BBB transport lactate and pyruvate, short chain monocarboxylic acids such as acetate, and ketone bodies such as γ-hydroxybutyrate and acetoacetate, which are essential for brain metabolism (4). Some types of acidic drug bearing a monocarboxylic acid moiety can cross the BBB via the monocarboxylic acid transport system(s). Use of the in vivo carotid artery injection technique and in vitro primary-cultured bovine BCECs disclosed a significant competitive inhibitory effect of salicylic acid and valproic acid on the transport of [^{3}H]acetic acid, whereas di- and tricarboxylic acids, amino acids, and choline were not inhibitory (36, 37). Pharmacologically active forms of 3-hydroxy-3-methylglutaryl coenzyme A (HMG-CoA) reductase inhibitors such as lipophilic [^{14}C]simvastatin acid (the most lipophilic derivative), lovastatin acid, and hydrophilic pravastatin, all of which contain a carboxylic acid moiety, are transported by proton/monocarboxylate cotransporter(s) at the BBB (38, 39). As shown in Table 1, less lipophilic pravastatin showed permeability comparable to that of [^{14}C]sucrose (39) and had a low affinity for the transporter responsible for the uptake of simvastatin acid, as evaluated from its inhibitory effect on the uptake of [^{14}C]simvastatin acid by primary-cultured BCECs (38). Simvastatin, a prodrug of simvastatin acid, is known to cause sleep disturbance, whereas pravastatin apparently does not, which suggests that these drugs may differ in ability to

Table 1 Effect of Unlabeled Simvastatin Acid on the
In Vivo Cerebrovascular Permeability Coefficient of
[^{14}C]HMG-CoA Reductase Inhibitors and [^{14}C]Sucrose
in Rats

	Permeability coefficient ($\mu l \ min^{-1}cm^{-2}$)[a]
[^{14}C]Sucrose	0.175 ± 0.011
[^{14}C]Sucrose + 1 mM simvastatin acid	0.150 ± 0.17
[^{14}C]Pravastatin	0.178 ± 0.021
[^{14}C]Lovastatin acid	0.442 ± 0.024
[^{14}C]Simvastatin acid	0.653 ± 0.058
[^{14}C]Simvastatin acid + 1 mM simvastatin acid	0.316 ± 0.095*
[^{14}C]Simvastatin	27.7 ± 3.9
[^{14}C]Lovastatin	57.9 ± 9.6

[a] Each value represents the mean ± SE of five experiments.
* Significant inhibition, $p < 0.05$. Residual radioactivity in cerebral
  blood vessels was not corrected.
Source: Ref. 38.

permeate through the BBB (see Table 1). The absence of CNS side effects of
pravastatin may be ascribed to its very low affinity for the monocarboxylate
transporter. Therefore, a strategy to avoid undesirable CNS side effects may
be to reduce the affinity of drugs for the transporter(s) functioning at the BBB.

MCT1, a proton-coupled monocarboxylate transporter that transports organic monocarboxylates such as benzoic acid and salicylic acid, has been
cloned from the rat intestine. By RT-PCR analysis, MCT1 was identified at
the blood–brain barrier as well as in various tissues (40). Recently, MCT1
was found on both the luminal and abluminal membranes of brain capillary
endothelial cells (41). These results suggest that MCT1 at the BBB functions
for the bidirectional transport of lactic acid and other monocarboxylate compounds. Since a Na^+/H^+ exchanger exists at the luminal membrane of the
brain capillary endothelial cells (42), protons are expected to be supplied by
this exchanger, providing a driving force to MCT1 for the enhanced transport
of monocarboxylates across the BBB from plasma to brain extracellular fluid.
The abundance of MCT1 in cerebral microvessels of suckling rats suggests
an important role of this transporter in the delivery of energy substrates to the

neonatal brain. Under normal physiological conditions in adult rats, however, this MCT1 presumably facilitates efflux of lactate, which is produced from glucose in the brain, from the brain into blood (41).

Although valproic acid, a monocarboxylate, inhibits the BBB transport of lactate, pyruvate, and acetate, it has been demonstrated, by means of an in situ brain perfusion technique in the rat, that the uptake of valproic acid into the brain across the BBB is not inhibited by coperfusion of short chain ($<C_4$) fatty acids. The uptake of [^{14}C]valproic acid was inhibited significantly by medium chain (C_6–C_{12}) fatty acids, suggesting that the monocarboxylate transporter is not involved in the BBB transport of valproic acid. Para-aminohippuric acid (PAH) inhibited the transport of valproic acid, whereas both cis- and trans-presence of medium chain dicarboxylates markedly stimulated the uptake of radiolabeled valproic acid (43). These observations suggest that the putative valproic acid transporter at the BBB may be an anion exchanger that operates in a manner similar to that reported for the PAH transporter OAT1, expressed at the basolateral membrane of the renal tubular epithelium (44). OAT2, cloned as a homologue of OAT1, is exclusively expressed in the liver and kidney (45) and transports organic anions such as salicylic acid, acetylsalicylate, PGE2, dicarboxylate, and PAH. An OAT family member that mediates the transport of valproic acid and PAH in an anion exchange manner may exist at the BBB.

B. Amine Transporters

The transport mechanisms for amine drugs have not yet been well elucidated, but passive diffusion and participation of carrier-mediated transport have been suggested for several drugs (4, 7, 8, 10, 46). An endogenous hydrophilic amine, choline, has been demonstrated to be taken up via a carrier-mediated transport mechanism (4, 7, 47, 48). When evaluated by the BUI method, uptake of [^{3}H]choline was inhibited by amine compounds (eperisone, scopolamine, thiamine, isoproterenol, and hemicholinium-3), whereas zwitterionic or anionic compounds were not inhibitory (48).

The observations previously reported for the BBB transport of amine compounds suggest that there are at least two different carrier-mediated transport mechanisms specific to choline and amine drugs. For example, carrier-mediated transport of H_1 antagonists was demonstrated (49, 50). Clarifying the BBB transport mechanism for H_1 antagonists is important, since H_1 antagonists often exhibit a significant sedative side effect, presumably caused by H_1 receptor blockade in the CNS. Uptake and transport studies making use of monolayers of primary-cultured BCECs and the carotid injection technique revealed saturable uptake into the brain of a classical H_1 antagonist, [^{3}H]me-pyramine (49, 50). The uptake was inhibited by amine drugs such as chlor-phenyramine and diphenhydramine, but not by choline, hemicholinium-3, or anionic drugs. Several H_1 antagonists—azelastine, ketotifen, cyproheptadine, emedastine, and cetirizine—competitively inhibited the uptake of [^{3}H]mepyr-amine by monolayers of primary-cultured BCECs to 8.6, 15.1, 15.8, 28.5, and 75.1% of the control, respectively, which suggests that they share common transport mechanisms with mepyramine (51). Among them, the weakest inhib-itory effect was observed with cetirizine, which has a carboxylated side chain. Accordingly, introduction of an anionic moiety within the molecule may de-crease affinity for the transporters, which would be desirable for drugs with CNS side effects.

C. Other Transport Systems Responsible for Drug Transport at the Blood–Brain Barrier

Transport systems at the BBB include a neutral amino acid transport system for phenylalanine, leucine, and other neutral amino acids, an acidic amino acid transport system for glutamate and aspartate, a basic amino acid transport system for arginine and lysine, and a β-amino acid transport system for β-alanine (4, 52). The design of amino acid analogues could, therefore, be a useful approach in developing effective CNS drugs. Several amino acid mi-metic drugs, such as L-dopa, α-methyldopa, α-methyltrypsin, baclofen, gaba-pentin (neurotin), and phenylalanine mustard, are expected to be taken up by the neutral amino acid transport system (4, 10). 6-[^{18}F]Fluoro-L-dopa (FDOPA) has been used to measure the central dopaminergic function in many species, including humans and monkeys. When the BBB transport of FDOPA was measured in adult monkeys by positron emission tomography scans, it was found to follow Michaelis–Menten kinetics and to be subject to competitive inhibition by the plasma large neutral amino acids (53). Orally administered and endogenous taurine is taken up by the β-alanine transporter at both luminal and abluminal membranes of brain capillary endothelial cells in a Na^+- and Cl^--dependent manner (54).

A transport study using primary cultured porcine BCECs suggested the existence of a Na^+-independent and saturable ($K_m = 28$ mM) transport system for L-carnitine (55). The proposed luminal transporter for L-carnitine seems to be different from the carnitine transporter *OCTN2* cloned recently in our laboratory, because OCTN2 takes up L-carnitine in a Na^+-dependent manner and exists abundantly in kidney, skeletal muscle, heart, and placenta, but not in brain (56). Currently there is much interest in the pharmacological treatment of patients with Alzheimer's disease, and acyl-L-carnitines are one group of candidate drugs. However, it is unclear whether acyl-L-carnitine is transported across the BBB via a saturable system or by passive diffusion.

Utilization of the hexose transport system for glucose and mannose and the nucleoside transport system for purine bases such as adenine and guanine, but not pyrimidine, is also an attractive strategy for obtaining the efficient delivery of drugs by means of an appropriate chemical modification so that they can be recognized and transported via these transporters (4, 10). The brain-type hexose transporter GLUT-1 has a very large V_{max} value. Some L-serinyl-β-D-glycoside analogues of Met[5]enkephalin have been shown to be transported across the BBB and produced a marked and long-lasting analgesia after intraperitoneal administration in mice (57). This result implies that GLUT-1 is responsible for transporting these glycopeptides into the CNS and indicates that glycosylation might be a promising way to deliver into the brain drugs with CNS activity but low permeability across the BBB.

Although it has not been established whether the Na^+-dependent hexose transporter SGLT is expressed at the BBB, a recent report suggested a participation of SGLT in the BBB transport of cycasin (58). Cycasin, methylazoxy-methanol-D-glucoside, is proposed to be a significant etiologic factor for the prototypical neurodegenerative disorder Western Pacific amyotrophic lateral sclerosis and for Parkinsonism–dementia complex. Cycasin is taken up into primary-cultured bovine BCECs in a dose-dependent manner with maximal uptake at a concentration of 10 μM. Since cycasin uptake was significantly inhibited by α-methyl-D-glucoside, a specific analogue for the Na^+-dependent glucose transporter, SGLT, as well as by phlorizin (a SGLT inhibitor), replacement of extracellular NaCl with LiCl, and dinitrophenol (an inhibitor of energy metabolism), cycasin is suggested to be transported across the BBB via a Na^+ energy-dependent SGLT (58).

IV. TRANSPORT OF PEPTIDES ACROSS THE BLOOD–BRAIN BARRIER

There are many biologically active peptides in the CNS, so it is important to develop delivery systems for neuropharmaceutical peptides across the BBB. Since peptides are in general relatively large, hydrophilic, and unstable, efficient permeation into the brain cannot be expected unless a specific delivery strategy is employed. Such strategies for peptides have been well documented and reviewed (3–7, 10, 59–61). One approach is a physiologically based strategy involving the use of carrier-mediated transport mechanisms for relatively small di- or tripeptides, and adsorptive-mediated endocytosis (AME) or receptor-mediated endocytosis (RME) for larger peptides. Another approach is the pharmacologically based strategy of converting water-soluble peptides into lipid-soluble ones.

The physiologically based strategy to manipulate large peptides to enhance their permeation into the brain involves the synthesis of chimeric peptides. These are formed by the covalent attachment of a nonpermeant but pharmacologically effective peptide to an appropriate vector, which can drive transport across the BBB. Pardridge (59, 60) proposed a novel strategy for the delivery of chimeric peptides through the BBB. First, the chimeric peptide is transported into brain endothelial cytoplasm by AME or RME. Second, the intact chimeric peptide is transferred into the brain interstitial space by adsorptive-mediated or receptor-mediated exocytosis. Third, the linkage between the vector and the pharmacologically active peptide is cleaved. Finally, the released peptide exerts its pharmacological effect in the CNS.

Examples of delivery of peptides into the brain by means of the above-mentioned strategies are described in Secs. A–C.

A. Carrier-Mediated Transport of Peptides

Oligopeptide transporters PepT1 and PepT2, cloned from rat, rabbit, and human, are highly expressed in the intestine and kidney, respectively (62). PepT2 is also expressed in liver and brain. Both transporters deliver orally effective β-lactam antibiotics, angiotensin-converting enzyme (ACE) inhibitors, bestatin, and other peptide-mimetic drugs (62). Another oligopeptide transporter, HPT, responsible for transport of histidine oligopeptides and histidine itself, has been cloned from brain (63). It has not been clarified yet whether PepT2 and HPT exist at the BBB.

Some transport systems other than PepT1, PepT2, and HPT for small peptides have been suggested to exist at the BBB (5, 6). Enkephalins were demonstrated to be taken up efficiently into the brain, and the involvement of a saturable mechanism for the transport of leucine enkephalin was shown by the in situ vascular brain perfusion technique. Several other peptides, thyrotropin-releasing hormone, arginine–vasopressin, peptide-T, α-melanocyte-stimulating hormone, luteinizing-hormone-releasing hormone, δ-sleep-inducing peptide, and interleukin 1, have been shown to cross the BBB, and some of them showed saturable uptake, suggesting participation of carrier-mediated transport. The peptide transport mechanisms involved are classified as peptide transport systems PTS-1 to PTS-5 (5, 6), although these transporter proteins have not been isolated as yet. It was recently indicated that [D-penicillamine-

2,5]enkephalin penetrates the BBB via a combination of diffusion and saturable transport, and biphalin ([Tyr-D-Ala-Gly-Phe-NH]$_2$) does so via diffusion and the large neutral amino acid carrier (64).

When the transport of the reduced form of glutathione across the BBB was examined by using the BUI method, the uptake of ^{35}S-labeled GSH was found to be saturable, with a K_m value of 5.8 mM and the surprisingly high maximal extraction of 30% (65). A functional expression study in *Xenopus laevis* oocytes of RNA obtained from bovine brain capillary revealed a Na$^+$-dependent GSH transporter as a novel BBB transporter distinct from the rat canalicular GSH transporter or γ-glutamyltranspeptidase (66).

Whether these peptide transporters existing at the BBB can be utilized for the specific brain delivery of small peptides or peptide-mimetic drugs remains to be fully investigated.

B. Adsorptive-Mediated Endocytosis

Utilization of the transcytosis mechanism for enhancement of the brain delivery of peptides seems promising because of its applicability to a wide range of peptides, including synthetic ones. AME is triggered by an electrostatic interaction between a positively charged moiety of the peptide and a negatively charged plasma membrane surface region. AME has lower affinity and higher capacity than RME, and these properties should be favorable for delivery of peptides to the brain (7, 59–61).

Several studies have been done on AME of neuropharmaceutical peptides, which have the characteristics of stability to enzymes and cationic charge as a consequence of suitable chemical modifications of the native peptides. Ebiratide (H-Met(O$_2$)-Glu-His-Phe-D-Lys-Phe-HMeThy-Arg-MeArg-D-Leu-NH(CH$_2$)$_8$NH$_2$), a synthetic peptide analogous to adrenocorticotropic hormone that is used to treat Alzheimer's disease, is positively charged with an isoelectric point of 10, and its resistance to metabolism has been enhanced by chemical modifications of the constituent natural amino acids. The internalization of [^{125}I]ebiratide was saturable in primary cultures of bovine BCECs. Furthermore, the characteristics of its internalization were consistent with AME in various respects, including energy dependence and the inhibitory effects of polycationic peptides and endocytosis inhibitors (7, 61, 67). To prove the transcytosis of ebiratide across the BBB, a capillary depletion study and brain microdialysis study were performed (68). After infusion of [^{125}I]ebiratide or [^{14}C]sucrose into the internal carotid artery for 10 minutes, the rat brain hemisphere was isolated and treated by the capillary depletion method. Since ebiratide is metabolized in vivo, HPLC analysis was performed to determine the unmetabolized ebiratide in the brain parenchyma and capillary fractions. The apparent distribution volume $V_{d\,app}$ of unmetabolized ebiratide in the brain parenchyma fraction (167.8 ± 62.2 µl/g brain) was about sevenfold greater than that of sucrose (24.9 ± 4.0 µl/g brain) and was also 35-fold greater than that of sucrose in the brain capillary fraction (6.2 ± 1.8 µl/g brain). As shown in Fig. 3, the unmetabolized ebiratide determined by HPCL analysis of the dialysate in the brain microdialysis accounted for more than 80% of total radioactivity, indicating that ebiratide crosses the BBB in an intact form (69). Similar AME and transcytosis at the BBB have been demonstrated for E-2078 (H-MeTyr-Gly-Gly-Phe-Leu-Arg-MeArg-D-Leu-NHC$_2$H$_5$), an analogue of dynorphin$_{1-8}$, which has high affinity for the opioid receptor and exhibits an analgesic activity after systemic administration (61, 70, 71).

The structural specificity of AME at the BBB was evaluated by the use of synthesized model peptides 001-C8, H-MeTyr-Arg-MeArg-D-Leu-NH(CH$_2$)$_8$NH$_2$, with primary-cultured bovine BCECs. Upon comparison of uptakes of peptides modified with 1,8-octanediamine, 1,5-pentanediamine, 1,2-ethanediamine or ethylamide, and peptides with a free carboxyl terminal against that of [^{3}H]PEG900, it was concluded that the C-terminal structure

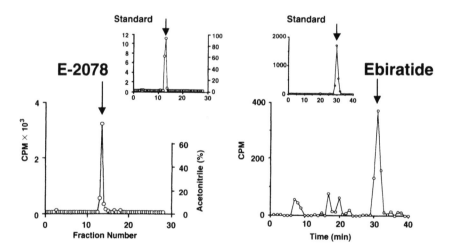

Fig. 3 HPLC elution profiles for [^{125}I]E-2078 and [^{125}I]Ebiratide in the dialysate during the infusion into the rat internal carotid artery of a solution containing each at 5 µCi/ml for 10 minutes at a rate of 1.0 or 0.5 ml/min. Inserts show HPLC elution profiles of standard samples of [^{125}I]E-2078 and [^{125}I]Ebiratide. (From Refs. 68 and 70.)

and basicity of the peptides are the most important determinants of uptake by AME at the BBB, not the number of constituent amino acids of peptides (72, 73). Other large molecules that penetrate the BBB via AME include various polycationic proteins such as β-endorphin-cationized albumin complex (74), histone (75), and avidin (76).

C. Receptor-Mediated Endocytosis

Transferrin receptor is present at a relatively high concentration on the vascular endothelium of the brain capillaries. The OX-26 antibody, which is a mouse IgG2a monoclonal antibody to rat transferrin receptor, binds to an extracellular epitope on the transferrin receptor that is distinct from the transferrin ligand binding site, so binding of the OX-26 monoclonal antibody to the receptor does not interfere with transferrin binding. By means of the capillary depletion technique for estimating the extent of transport into brain parenchyma, it was established that the OX-26 monoclonal antibody is transported across the BBB and is effective as a drug delivery vehicle (59, 60). Saito et al. (77) used OX-26 monoclonal antibody to enhance the BBB permeation of β-amyloid peptide, A-β_{1-40}, which binds to preexisting amyloid plaques in Alzheimer's disease. Since the dementia in Alzheimer's disease is correlated with amyloid deposition in the brain, if the peptide could be delivered to the brain, we would have a useful tool for the diagnostic assay of Alzheimer's disease.

The utilization of insulin receptor existing at the BBB is also effective. When human insulin was infused into mice for 48 hours subcutaneously at several doses, both brain and blood concentrations measured by species-specific radioimmunoassay increased with increasing dose (78). Since the mouse cannot make human insulin, blood was the only source for the human insulin in the brain, indicating that insulin does indeed cross the BBB. The relationship between the concentration of human insulin in brain and blood was nonlinear, showing that BBB passage of human insulin occurs via a saturable transport system, although it is unclear whether the insulin receptor mediates this transport (78). There is clear evidence for receptor-mediated delivery of an insulin conjugate and its degradation fragments. After mice received an intravenous injection of horseradish peroxidase (HRP, MW 40,000) conjugated with insulin, as shown in Fig. 4, the HRP activity in the brain was significantly higher than it was after HRP injection (79). A fragment of insulin obtained by trypsin digestion showed high insulin receptor binding activity and scarcely any hypoglycemic activity in mice, suggesting that this fragment may be useful as a carrier to transport therapeutic peptides across the BBB via RME (79).

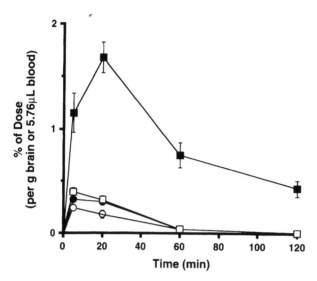

Fig. 4 Time courses of horseradish peroxidase (HRP) activity in the brain (squares) and blood (circles) of mice after intravenous administration of HRP (230 µg; open symbols) and insulin–HRP conjugate (213 µg; solid symbols). HRP activity in blood represents the results obtained in 5.76 µl of blood. (From Ref. 79.)

V. EFFLUX TRANSPORT SYSTEMS FOR DRUGS

Before 1992, most investigations on drug delivery to the brain were primarily focused on the influx mechanism from circulating blood into the brain. The concept of the BBB was changed by recent findings showing that efflux transport systems expressed at the luminal and/or abluminal membranes of BCECs act to restrict the accumulation of some drugs. As described earlier in this chapter, Pgp is one such efflux transporter. These lines of study are likely to lead to the discovery of novel BBB efflux transporters. Furthermore, it should become possible to enhance the efficacy of drugs or to reduce side effects in the CNS by modulating the function of Pgp and the other efflux transporters to respectively increase or decrease the apparent BBB permeability to the drugs.

Two ATP-binding cassette (ABC) transporters acting as energy-dependent efflux pumps, MDR-encoded Pgp, and the recently described multidrug resistance–associated protein (MRP) are known to operate as drug efflux transport systems. Pgp actively transports many cytotoxic hydrophobic drugs, while MRP transports organic acids and selected glucuronide or glutathione conjugates out of the cells. In the brain, Pgp is expressed highly in the capillary endothelial cells forming the BBB. Recently MRP was confirmed by Western blotting and RT-PCR analysis to be expressed in bovine BCECs, implying that MRP at the BBB may also have an important role in limiting the exposure of the brain to many endogenous and exogenous compounds, including both toxic and therapeutic compounds (80). Sugiyama and his coauthors (81) indicated that the ATP-dependent uptake of 2,4-dinitrophenyl-S-glutathione (DNP-SG) and leukotriene C4 (LTC4) into membrane vesicles prepared from mouse BCECs (MBEC4) was saturable, with K_m values of 0.56 μM for DNP-SG and 0.22 μM for LTC4. Northern blot and Western blot analyses showed the expression of murine MRP1, but not cMOAT (MRP2). This result suggests that MRP1 is responsible for the unidirectional, energy-dependent efflux of organic anions such as DNP-SG and LTC4 from the brain into the circulating blood across the BBB (81). However, an another recent study indicated that in rat brain, MRP1, a member of the MRP family, is not predominantly expressed at the BBB but may be present in other cerebral cells (82). The function and localization of MRP in the brain remain to be characterized.

There may be an another ATP-dependent efflux pump at the BBB. The unidirectional brain-to-blood transport system for corticotropin-releasing hormone (CRH) across the BBB could be instrumental in the homeostasis of central CRH. Intracerebroventricularly injected [^{125}I]CRH was rapidly transported out of the brain with a half-time ($t_{1/2}$) of 15 minutes, much faster than that of albumin ($t_{1/2} = 50$ min). The efflux transport was saturable and was inhibited by tumor necrosis factor α and β-endorphin, verapamil, ouabain, and colchicine, but not by cyclosporin A, whereas the transport was increased by corticosteronein (83). These results suggest that the specific unidirectional brain-to-blood transport system for CRH is dependent on energy and calcium channels, involves microtubules, is independent of the Pgp transporter, and is acutely modulated by adrenal steroids, cytokines, and endogenous opiates.

The brain efflux index method also demonstrated the efflux from brain to blood of PAH, taurocholic acid, valproic acid, and so on. The efflux of [^{3}H]taurocholic acid was saturable and was inhibited by cholic acid, a cationic cyclic octapeptide (octreotide, a somatostatin analogue), an anionic cyclic pentapeptide (BQ-123, an endothelin receptor antagonist), and probenecid, but not PAH. Although efflux of [^{3}H]BQ-123 from the brain was confirmed, no significant efflux of [^{14}C]octreotide was observed. Since the mutual inhibition between taurocholic acid and BQ-123 was noncompetitive, these results suggest that taurocholic acid and BQ-123 are transported from the brain to the circulating blood across the BBB via different transport systems (84).

Salicylate, benzoate, and probenecid were suggested to be transferred via the monocarboxylic acid transport system from the brain interstitial fluid to plasma across the blood–brain barrier (85). The restricted distribution of probenecid in the brain may thus be ascribed to efficient efflux from the brain via MCT1 or the related MCT2, which has been confirmed recently to be expressed at the BBB (86); other organic anion transporters also are known.

Participation of MCT1 in the efflux of metals from the brain to blood was suggested (87). When unbound extracellular aluminum was measured by microdialysis in the jugular vein as well as in the right and left frontal cortices of rats, the steady state brain-to-blood ratio ($K_{p,brain}$) was less than 1, suggesting the presence of a process other than diffusion for transfer across the BBB. After alkalinizing dialysis via a brain microdialysis probe (to pH 10.2), the $K_{p,brain}$ value increased from 0.35 to 0.80. The addition to the brain dialysate of a proton ionophore, p-(trifluoromethoxy)phenylhydrazone, increased the $K_{p,brain}$ value from 0.21 to 0.61. These increased $K_{p,brain}$ values suggest that a proton-dependent process is removing aluminum from brain extracellular fluid. Addition of mersalyl acid, which inhibits MCT1 but not MCT2, to the brain dialysate increased the $K_{p,brain}$ value from 0.19 to 0.87 (87). Since the monocarboxylate transporters are the only known proton-dependent transporters at the BBB, MCT1 is suggested to mediate at least partially the efflux of aluminum from brain extracellular fluid.

Very recently, the organic anion transporters $oatp2$ and $oatp3$ were cloned by Abe et al. (88) from rat retina as homologues of the Na^+-independent multispecific organic anion transporter $oatp1$, which was isolated from rat liver (89). Also, Noe et al. cloned $oatp2$ from rat brain (90). Although oatp1 is not expressed in the brain, oatp2 is exclusively expressed in the liver, brain, and retina, and it transports thyroid hormones T3 and T4, as well as taurocholate (88), bile acids, estrogen conjugates, ouabain, and digoxin (90). Noe et al. suggested that oatp2 plays a key role in the brain accumulation and toxicity of digoxin (90). However, it remains to be clarified whether oatp2 is expressed at the BBB, and if so, whether it operates for uptake of the foregoing transportable compounds from the blood into the brain or for their efflux from the brain into blood.

VI. PERSPECTIVES FOR DRUG DEVELOPMENT AND CONCLUSION

Recent advances in studies on the BBB transport of xenobiotics, as well as nutrients and neuroactive agents, have led to a change in the classical concept of the blood–brain barrier. As described above, the BBB acts as a dynamic interface that has physiological functions of specific and selective uptake from blood to brain and efflux from brain to blood for many compounds. Early work on drug delivery to the brain was based on the strategy of adding lipophilicity to candidate drugs, on the assumption that transport of drugs across the lipoidal membrane barrier of BCECs, which form the BBB, is diffusion-limited. However, this strategy has often resulted in a poor oral bioavailability due to a poor dissolution rate in the gastrointestinal lumen, extensive first-pass metabolism in the intestine and/or liver, or secretion from the intestinal epithelial cells and into bile across the bile canalicular membrane by multidrug transporters such as P-glycoprotein and cMOAT, or alternatively, easy penetration from the circulation into the brain, resulting in severe side effects in the CNS. Accordingly, for the development of brain-specific drug delivery systems for neuroactive drugs, a better strategy may be to utilize the specific transport mechanisms at the BBB.

There is at least one successful example of tissue-selective drug delivery by utilizing specialized transporter(s) expressed in the targeted tissue cells. Many derivatives of new quinolone antibacterial agents are known to cause severe CNS side effects, such as convulsion. We have obtained evidence of active efflux of lipophilic quinolones by Pgp at the BBB. This finding implies

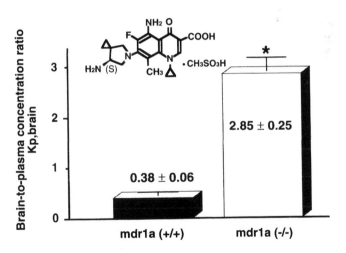

Fig. 5 Comparison of brain distribution of HSR-903 between normal mice and mice in which the *mdr1a* gene had been disrupted after intravenous doses of 13 mg/kg. The brain-to-plasma concentration ratio $K_{p,brain}$ at the steady state is shown as the mean ± SEM ($n = 3$). Asterisk indicates the significant difference versus the $K_{p,brain}$ value in normal mice, $p < 0.05$. (From Tsuji, unpublished observations.)

that development of new quinolones with a higher affinity for Pgp minimize the CNS side effects. A newly synthesized quinolone, HSR-903, exhibits relatively high lipophilicity, with the octanol–water (pH 7.4) partition coefficient of 2.58. HSR-903 is well absorbed from the intestine, probably via a carrier-mediated system, is well distributed into the lung via a Na^+- and Cl^--dependent transport system, shows low penetration across the BBB owing to the active efflux by Pgp (see Fig. 5) (unpublished observation), and exhibits a high biliary secretion as a result of carrier-mediated uptake by hepatocytes and transport by cMOAT (91, 92).

Therefore, a detailed understanding of the uptake and efflux mechanisms at the BBB and other tissues, including intestine, would be very helpful for targeting drugs to the brain to provide the expected CNS pharmacological effect, or for reducing the BBB penetration of drugs to minimize unwanted side effects in the CNS.

ACKNOWLEDGMENTS

I thank Drs. Ikumi Tamai, Tetsuya Terasaki, and Y.-S. Kang for their contribution to studies in this laboratory on the BBB transport of drugs. This research was supported in part by grants in-aid for scientific research from the Japanese Ministry of Education, Science, Sports and Culture, a grant from the Japan Research Foundation for Clinical Pharmacology, and a grant from the Tokyo Biochemical Research Foundation.

REFERENCES

1. EM Cornford. The blood–brain barrier, a dynamic regulatory interface. Mol Physiol 7:219–260, 1985.
2. VA Levin. Relationship of octanol/water partition coefficient and molecular weight to rat brain capillary permeability. J Med Chem 23:682–684, 1980.
3. WM Pardridge, D Triguero, J Yang, PA Cancilla. Comparison of in vitro and in vivo models of drug transcytosis through the blood–brain barrier. J Pharmacol Exp Ther 253:884–891, 1990.
4. WM Pardridge. Transport of small molecules through the blood–brain barrier: Biology and methodology. Adv Drug Delivery Rev 15:5–36, 1995.
5. BV Zlokovic. Cerebrovascular permeability to peptides: Manipulations of transport systems at the blood–brain barrier. Pharm Res 12:1395–1406, 1995.
6. WA Banks, AJ Kastin, CM Barrer. Delivering peptides to the central nervous system: Dilemmas and strategies. Pharm Res 8:1345–1350, 1991.

7. I Tamai, A Tsuji. Drug delivery through the blood–brain barrier. Adv Drug Delivery Rev 19:401–424, 1996.

8. A Tsuji, I Tamai. Blood–brain barrier function of P-glycoprotein. Adv Drug Delivery Rev 25:287–298, 1997.

9. AH Schinkel. P-Glycoprotein, a gatekeeper in the blood–brain barrier. Adv Drug Delivery Rev 36:179–194, 1999.

10. DJ Begley. The blood–brain barrier: Principles for targeting peptides and drugs to the central nervous system. J Pharm Pharmacol 48:136–146, 1996.

11. A Tsuji, T Terasaki, Y Takabatake, Y Tenda, I Tamai, T Yamashima, S Moritani, T Tsuruo, J Yamashita. P-Glycoprotein as the drug efflux pump in primary cultured brain capillary endothelial cells. Life Sci 51:1427–1437, 1992.

12. A Tsuji, I Tamai, A Sakata, Y Tenda, T Terasaki. Restricted transport of cyclosporin A across the blood–brain barrier by a multidrug transporter, P-glycoprotein. Biochem Pharmacol 46:1096–1099, 1993.

13. A Sakata, I Tamai, K Kawazu, Y Deguchi, T Ohnishi, A Saheki, A Tsuji. In vivo evidence for ATP-dependent and P-glycoprotein-mediated transport of cyclosporin A at the blood–brain barrier. Biochem Pharmacol 48:1989–1992, 1994.

14. T Ohnishi, I Tamai, K Sakanaka, A Sakata, T Yamashima, J Yamashita, A Tsuji. In vivo and in vitro evidence for ATP-dependency of P-glycoprotein-mediated efflux of doxorubicin at the blood–brain barrier. Biochem Pharmacol 49:1541–1544, 1995.

15. T Tatsuta, M Naito, T Oh-hara, I Sugawara, T Tsuruo. Functional involvement of P-glycoprotein in blood–brain barrier. J Biol Chem 267:20383–20391, 1992.

16. J Greenwood. Characterization of a rat retinal endothelial cell culture and the expression of P-glycoprotein in brain and retinal endothelium in vitro. J Neuroimmunol 39:123–132, 1992.

17. EJ Hegmann, HC Bauer, RS Kerbel. Expression and functional activity of P-glycoprotein in cultured cerebral capillary endothelial cells. Cancer Res 52:6969–6975, 1992.

18. Q Wang, H Yang, DW Miller, WF Elmquist. Effect of the P-glycoprotein inhibitor, cyclosporin A, on the distribution of rhodamine-123 to the brain: An in vivo microdialysis study in freely moving rats. Biochem Biophys Res Commun 211:719–726, 1995.

19. EC de Lange, G de Bock, AH Schinkel, AG de Boer, DD Breimer. BBB transport and P-glycoprotein functionality using *mdr1a* (−/−) and wild-type mice. Total brain versus microdialysis concentration profiles of rhodamine-123. Pharm Res 15:1657–1665, 1998.

20. AH Schinkel, JJM Smit, O van Tellingen, JH Beijnen, E Wagenaar, L van Deemter, CAAM Mol, MA van der Valk, EC Robanus-Maandag, HPJ te Riele, AJM Berns, P Borst. Disruption of the mouse mdr1a P-glycoprotein gene leads to a deficiency in the blood–brain barrier and to increased sensitivity to drugs. Cell 77:491–502, 1994.

21. AH Schinkel, U Mayer, E Wagenaar, CAAM Mol, JJM Smit, MA van der Valk,

AC Voordouw, H Spits, O van Tellingen, JM Zijlmans, WE Fibbe, P Borst. Normal viability and altered pharmacokinetics in mice lacking mdr1-type (drug-transporting) P-glycoproteins. Proc Natl Acad Sci USA 94:4028–4033, 1997.

22. AH Schinkel, E Wagenaar, L van Deemter, CAAM Mol, P Borst. Absence of the mdr1a P-glycoprotein in mice affects tissue distribution and pharmacokinetics of dexamethasone, digoxin, and cyclosporin A. J Clin Invest 96:1698–1705, 1995.

23. U Mayer, E Wagenaar, JH Beijnen, JW Smit, DKF Meijer, J van Asperen, P Borst, AH Schinkel. Substantial excretion of digoxin via the intestinal mucosa and prevention of long-term digoxin accumulation in the brain by the mdr1a P-glycoprotein. Br J Pharmacol 119:1038–1044, 1996.

24. AH Schinkel, E Wagenaar, CAAM Mol, L van Deemter, L. Deemter. P-glycoprotein in the blood–brain barrier of mice influences the brain penetration and pharmacological activity of many drugs. J Clin Invest 97:2517–2524, 1996.

25. U Mayer, E Wagenaar, B Dorobek, JH Beijnen, P Borst, AH Schinkel. Full blockade of intestinal P-glycoprotein and extensive inhibition of blood–brain barrier P-glycoprotein by oral treatment of mice with PSC833. J Clin Invest 100: 2430–2436, 1997.

26. RB Kim, MF Fromm, C Wandel, B Leake, AJJ Wood, DM Roden, GR Wilkinson. The drug transporter P-glycoprotein limits oral absorption and brain entry of HIV-1 protease inhibitors. J Clin Invest 101:289–294, 1998.

27. J van Asperen, U Mayer, O van Tellingen, JH Beijnen. The functional role of P-glycoprotein in the blood–brain barrier. J Pharm Sci 86:881–884, 1997.

28. EG Shuetz, K Yasuda, K Arimori, JD Schuetz. Human MDR1 and mouse mdr1a P-glycoprotein alter the cellular retention and disposition of erythromycin, but not retinoic acid or benzo(a)pyrene. Arch Biochem Biophys 350:340–347, 1998.

29. JF Watchko, MJ Daood, TW Hassen. Brain bilirubin content is increased in P-glycoprotein-deficient transgenic null mutant mice. Pediat Res 44:763–766, 1998.

30. T Yamashima, T Ohnishi, Y Nakajima, T Terasaki, M Tanaka, J Yamashita, T Sasaki, A Tsuji. Uptake of drugs and expression of P-glycoprotein in the rat 9L glioma. Exp Brain Res 95:41–50, 1993.

31. A Tsuji. P-Glycoprotein-mediated efflux transport of anticancer drugs at the blood–brain barrier. Ther Drug Monit 20:588–590, 1998.

32. WH Oldendorf. Measurement of brain uptake of radiolabeled substances using a tritiated water internal standard. Brain Res 24:372–376, 1996.

33. M Lemaire, A Bruelisauer, P Guntz, H Sato. Dose-dependent brain penetration of SDZ PSC833, a novel multidrug resistance–reversing cyclosporin, in rats. Cancer Chemother Pharmacol 38:481–486, 1996.

34. N Drion, M Lemaire, J-M Lefauconnier, JM Scherrmann. Role of P-glycoprotein in the blood–brain barrier transport of colchicine and vinblastine. J Neurochem 67:1688–1693, 1996.

35. S Desrayaud, P Guntz, J-M Schermann, M Lemaire. Effect of the P-glycoprotein inhibitor, SDZ PSC833, on the blood and brain pharmacokinetics of colchicine. Life Sci 61:153–163, 1997.

36. T Terasaki, S Takakuwa, S Moritani, A Tsuji. Transport of monocarboxylic acids at the blood–brain barrier: Studies with monolayers of primary cultured bovine brain capillary endothelial cells. J Pharmacol Exp Ther 258:932–937, 1991.

37. T Terasaki, Y-S Kang, T Ohnishi, A Tsuji. In vitro evidence for carrier-mediated uptake of acidic drugs by isolated bovine brain capillaries. J Pharm Pharmacol 43:172–176, 1991.

38. A Tsuji, A Saheki, I Tamai, T Terasaki. Transport mechanism of 3-hydroxy-3-methylglutaryl coenzyme A reductase inhibitors at the blood–brain barrier. J Pharmacol Exp Ther 267:1085–1090, 1993.

39. A Saheki, T Terasaki, I Tamai, A Tsuji. In vivo and in vitro blood–brain barrier transport of 3-hydroxy-3-methylglutaryl coenzyme A (HMG-CoA) reductase inhibitors. Pharm Res 11:305–311, 1994.

40. H Takanaga, I Tamai, S Inaba, Y Sai, H Higashida, H Yamamoto, A Tsuji. cDNA cloning and functional characterization of rat intestinal monocarboxylate transporter. Biochem Biophys Res Commun 217:370–377, 1995.

41. DZ Gerhart, BE Enerson, OY Zhdankina, RL Leino, LR Drewes. Expression of monocarboxylate transporter MCT1 by brain endothelium and glia in adult and suckling rats. Am J Physiol 273:E207–E213, 1997.

42. SR Ennis, XD Ren, AL Betz. Mechanisms of sodium transport at the blood–brain barrier studied with in situ perfusion of rat brain. J Neurochem 66:756–763, 1996.

43. KD Adkison, DD Shen. Uptake of valproic acid into rat brain is mediated by medium-chain fatty acid transporter. J Pharmacol Exp Therap 276:1189–1200 (1996).

44. T Sekine, N Watanabe, M Hosoyamada, Y Kanai, H Endou. Expression cloning and characterization of a novel multispecific organic anion transporter. J Biol Chem 272:18526–18529, 1997.

45. T Sekine, SH Cha, M Tsuda, N Aplwattanakul, N Nakajima, Y Kanai, H Endou. Identification of multispecific anion transporter 2 expressed predominantly in the liver. FEBS Lett 429:179–182, 1998.

46. F Joo. Endothelial cells of the brain and other organ systems: Some similarities and differences. Prog Neurobiol 48:255–273, 1996.

47. Y-S Kang, T Terasaki, A Tsuji. Dysfunction of choline transport system through the blood–brain barrier in stroke-prone spontaneously hypertensive rats. J Pharmacobio-Dyn 13:10–19, 1990.

48. Y-S Kang, T Terasaki, T Ohnishi, A Tsuji. In vivo and in vitro evidence for a common carrier-mediated transport of choline and basic drugs through the blood–brain barrier. J Pharmacobio-Dyn 13:353–360, 1990.

49. M Yamazaki, H Fukuoka, O Nagata, H Kato, Y Ito, T Terasaki, A Tsuji. Transport mechanism of an H_1-antagonist at the blood–brain barrier: Transport mechanism of mepyramine using the carotid injection technique. J Pharmacobio-Dyn 17:676–679, 1994.

50. M Yamazaki, T Terasaki, K Yoshioka, O Nagata, H Kato, Y Ito, A Tsuji. Mepyramine uptake into bovine brain capillary endothelial cells in primary monolayer cultures. Pharm Res 11:975–978, 1994.

51. M Yamazaki, T Terasaki, K Yoshioka, O Nagata, H Kato, Y Ito, A Tsuji. Carrier-mediated transport of H_1-antagonist at the blood–brain barrier: A common transport system of H_1-antagonists and lipophilic basic drugs. Pharm Res 11:1516–1518, 1994.
52. WM Pardridge. Blood–brain barrier carrier-mediated transport and brain metabolism of amino acids. Neurochem Res 23:635–644, 1998.
53. DB Stout, SC Huang, WP Melege, MJ Releigh, ME Phelps, JR Barrio. Effects of large neutral amino acid concentrations on 6-[^{18}F]fluoro-L-DOPA kinetics. J Cereb Blood Flow Metab 18:43–51, 1998.
54. I Tamai, M Senmaru, T Terasaki, A Tsuji. Na^+ and Cl^--dependent transport of taurine at the blood–brain barrier. Biochem Pharmacol 50:1783–1793, 1995.
55. E Mroczkowska, H-J Galla, MJ Nalecz, KA Nalecz. Evidence for an asymmetrical uptake of L-carnitine in the blood–brain barrier in vitro. Biochem Biophys Res Commun 241:127–131, 1997.
56. I Tamai, R Ohashi, J Nezu, H Yabuuchi, Oku A, Shimane M, Sai Y, Tsuji A. Molecular and functional identification of sodium ion-dependent high affinity human carnitine transporter OCTN2. J Biol Chem 273:20378–20382, 1998.
57. B Polt, F Porreca, LZ Szabo, EJ Bilsky, P Davis, TJ Abbruscato, TP Davis, R Harvath, HI Yamamura, VJ Hruby. Glycopeptide enkephalin analogues produce analgesia in mice: Evidence for penetration of the blood–brain barrier. Proc Natl Acad Sci USA 91:7114–7118, 1994.
58. T Matsuoka, T Nishizaki, GE Kisby. Na^+-dependent and phlorizin-inhibitable transport of glucose and cycasin in brain endothelial cells. J Neurochem 70:772–777, 1998.
59. WM Pardridge. Peptide Drug Delivery to the Brain. New York: Raven Press, 1991.
60. WM Pardridge. Blood–brain barrier peptide transport and peptide drug delivery to the brain. In: MD Taylor, GL Amidon, eds. Peptide-Based Drug Design. Washington, DC: American Chemical Society, 1995, pp 265–296.
61. T Terasaki, A Tsuji. Oligopeptide drug delivery to the brain, importance of absorptive-mediated endocytosis and P-glycoprotein associated active efflux transport at the blood-brain barrier. In: MD Taylor, GL Amidon, eds. Peptide-Based Drug Design. Washington, DC: American Chemical Society, 1995, pp 297–316.
62. A Tsuji, I Tamai. Carrier-mediated intestinal transport of drugs. Pharm Res 13:963–977, 1996.
63. T Yamashita, S Shimada, W Guo, K Sato, E Kohmura, T Hayakawa, T Takagi, M Tohyama. Cloning and functional expression of a brain peptide/histidine transporter. J Biol Chem 272:10205–10211, 1997.
64. RD Egleton, TJ Abbruscto, SA Thomas, TP Davis. Transport of opioids into the central nervous system. J Pharm Sci 87:1433–1439, 1998.
65. R Kannan, JF Kuhlenkamp, E Jeandidier, H Trinh, M Ookhtens, N Kaplowitz. Evidence for carrier-mediated transport of glutathione across the blood–brain barrier in the rat. J Clin Invest 82:2009–2013, 1990.
66. R Kannan, JR Yi, D Tang, Y Li, BV Zlokovic, N Kaplowitz. Evidence for the existence of a sodium-dependent glutathione (GSH) transporter. Expression of

bovine brain capillary mRNA and size fractions in *Xenopus laevis* oocytes and dissociation from gamma-glutamyltranspeptidase and facilitative GSH transporters. J Biol Chem 271:9754–9758, 1996.

67. T Terasaki, S Takakuwa, A Saheki, S Moritani, T Shimura, S Tabata, A Tsuji. Absorptive-mediated endocytosis of an adrenocorticotropic hormone (ACTH) analogue, Ebiratide, into the blood–brain barrier: Studies with monolayers of primary cultured bovine brain capillary endothelial cells. Pharm Res 9:529–534, 1992.

68. T Shimura, S Tabata, T Ohnishi, T Terasaki, A Tsuji. Transport mechanism of a new behaviorally highly potent adrenocorticotropic hormone (ACTH) analog, Ebiratide, through the blood–brain barrier. J Pharmacol Exp Ther 258:459–465, 1991.

69. T Shimura, S Tabata, T Terasaki, Y Deguchi, A Tsuji. In vivo blood–brain barrier transport of a novel adrenocorticotropic hormone analogue, Ebiratide, determined by brain microdialysis and capillary depletion methods. J Pharm Pharmacol 44:583–588, 1992.

70. T Terasaki, K Hirai, H Sato, Y-S Kang, A Tsuji. Absorptive-mediated endocytosis of a dynorphin-like analgesic peptide, E-2078, into the blood–brain barrier. J Pharmacol Exp Ther 251:351–357, 1989.

71. T Terasaki, Y Deguchi, H Sato, K Hirai, A Tsuji. In vivo transport of dynorphin-like analgesic peptide, E-2078, through the blood–brain barrier: An application of brain microdialysis. Pharm Res 8:815–820, 1991.

72. I Tamai, Y Sai, H Kobayashi, M Kamata, T Wakamiya, A Tsuji. Structure–internalization relationship for adsorptive-mediated endocytosis of basic peptides at the blood–brain barrier. J Pharmacol Exp Ther 280:410–415, 1997.

73. T Wakamiya, M Kamata, S Kusumoto, H Kobayashi, Y Sai, I Tamai, A Tsuji. Design and synthesis of peptides passing through the blood–brain barrier. Bull Chem Soc Jpn 71:699–709, 1998.

74. AK Kumagai, JB Eisenberg, WM Pardridge. Absorptive-mediated endocytosis of cationized albumin and a beta-endorphin-cationized albumin chimeric peptide by isolated brain capillaries. J Biol Chem 262:15214–15219, 1987.

75. WM Pardridge, D Triguero, JB Buciak. Transport of histone through the blood–brain barrier. J Pharmacol Exp Ther 251:521–526, 1989.

76. WM Pardridge, RJ Boado. Enhanced cellular uptake of biotinylated antisense oligonucleotide or peptide mediated by avidin, a cationic protein. FEBS Lett 288:30–32, 1991.

77. Y Saito, J Buciak, J Yang, WM Pardridge. Vector-mediated delivery of ^{125}I-labeled β-amyloid peptide A-β_{1-40} through the blood–brain barrier and binding to Alzheimer disease amyloid of the A-β_{1-40}/vector complex. Proc Natl Acad Sci USA 92:10227–10231, 1995.

78. WA Banks, JB Jaspan, AJ Kastin. Selective, physiological transport of insulin across the blood–brain barrier: Novel demonstration by species-specific radioimmunoassays. Peptides 18:1257–1262, 1997.

79. M Fukuta, H Okada, S Iinuma, S Yanai, H Toguchi. Insulin fragments as a carrier for peptide delivery across the blood–brain barrier. Pharm Res 11:1681–1688, 1994.

80. H Huai-Yunn, DT Secrest, KS Mark, D Carney, C Brandquist, WF Elmquist, DW Miller. Expression of multidrug resistance–associated protein (MRP) in brain microvessel endothelial cells. Biochem Biophy Res Commun 243:816–820, 1998.

81. H Kurihara, H Suzuki, M Naito, T Tsuruo, Y Sugiyama. Characterization of efflux transport of organic anions in a mouse brain capillary endothelial cell line. J Pharmacol Exp Ther 285:1260–1265, 1998.

82. A Regina, A Koman, M Piciotti, B EL Hafny, MS Center, R Bergmann, PO Couraud. Mrp1 multidrug resistance-associated protein and P-glycoprotein expression in rat brain microvessel endothelial cells. J Neurochem 71:705–715, 1998.

83. JM Martins, WA Banks, AJ Kastin. Acute modulation of active carrier-mediated brain-to-blood transport of corticotropin-releasing hormone. Am J Physiol 272: E312–E319, 1997.

84. T Kitazawa, T Terasaki, H Suzuki, A Kakee, Y Sugiyama. Effect of taurocholic acid across the blood–brain barrier: Interaction with cyclic peptides. J Pharmacol Exp Ther 286:890–895, 1998.

85. Y Deguchi, K Nozawa, S Yamada, Y Yokoyama, R Kimura. Quantitative evaluation of brain distribution and blood–brain barrier efflux transport of probenecid in rats by microdialysis: Possible involvement of the monocarboxylic acid transport system. J Pharmacol Exp Ther 280:551–560, 1997.

86. L Pellerin, G Pellegi, JL Martin, PJ Magistretti. Expression of monocarboxylate transporter mRNAs in mouse brain: Support for a distinct role of lactate as an energy substrate for the neonate vs. adult brain. Proc Natl Acad Sci USA 31: 3990–3995, 1998.

87. DC Ackley, RA Yokel. Aluminum transport out of brain extracellular fluid is proton dependent and inhibited by mersalyl acid, suggesting mediation by monocarboxylate transporter (MCT1). Toxicology 127:59–67, 1998.

88. T Abe, M Kakyo, H Sakagami, T Tokui, T Nishio, M Tanemoto, N Nomura, SC Hebert, S Matsuno, H Kondo, H Yawo. Molecular characterization and tissue distribution of a new organic anion transporter subtype (oatp3) that transports thyroid hormones and taurocholate and comparison with oatp3. J Biol Chem 273: 22395–22401, 1998.

89. E Jacquemin, B Hagenbuch, B Stieger, AW Wolkoff, PJ Meier. Expression cloning of a rat liver Na$^+$-independent organic anion transporter. Proc Natl Acad Sci USA 91:133–137, 1994.

90. B Noe, B Hagenbuch, B Stieger, PJ Meier. Isolation of a multispecific organic anion and cardiac glycoside transporter from rat brain. Proc Natl Acad Sci USA 94:10346–10350, 1998.

91. M Murata, I Tamai, Y Sai, O Nagata, H Kato, Y Sugiyama, A Tsuji. Hepatobiliary transport kinetics of HSR-903, a new quinolone antibacterial agent. Drug Metab Dispos 1999, 26:1113–1119, 1998.

92. M Murata, I Tamai, Y Sai, O Nagata, H Kato, A Tsuji. Carrier-mediated lung distribution of HSR-903, a new quinolone antibacterial agent. J Pharmacol Exp Ther 1999, 289:79–84, 1999.

9
Drug Metabolism in the Brain: Benefits and Risks

Alain Minn, Ramon D.S. El-Bachá, Claire Bayol-Denizot, Philippe Lagrange, and Funmilayo G. Suleman
Université Henri Poincaré-Nancy 1, CNRS, Vandoeuvre-lès-Nancy, France

Daniela Gradinaru
National Institute of Gerontology and Geriatrics, Ana Aslan, Bucharest, Romania

I. INTRODUCTION

The brain is protected against chemical assault by cerebral endothelial cell membrane systems, forming a blood–brain barrier (BBB) that prevents the influx of most polar molecules. However, some lipophilic potentially toxic drugs and environmental pollutants can reach the central nervous system (CNS). Therefore, other efficient mechanisms of protection are needed to protect the brain from chemical insult. It has been clearly established that xenobiotic metabolism is catalyzed by a series of enzymes located both in the brain parenchyma and at blood–brain interfaces. Moreover, efficient transport mechanisms resulting from multidrug resistance protein (MRP) and P-glycoprotein (Pgp) activities can export both xenobiotics and some of their polar metabolites from the brain to the blood of the cerebral circulation. Therefore, to protect the central nervous system from foreign molecules that might disturb its homeostasis or display some toxicity, three different components are required: (a) a *physical barrier*, preventing the entry of polar and high molecular weight molecules; (b) a *metabolic barrier*, resulting from the activity of oxidases, reductases, or conjugating enzyme systems; and (c) an *active, ATP-dependent barrier*, due to the activity of multidrug resistance–related transport systems. Drug metabolism results both in inactivation of potentially toxic or

145

pharmacologically active molecules and in the possible formation of either reactive metabolites or activated oxygen species. As a consequence, some drug metabolism enzymatic activities may also trigger toxic events. The risk resulting from the formation of reactive metabolites may endanger brain functioning, inasmuch as the central nervous system possesses only a few scavenger systems, and because injured neurons do not regenerate. Moreover, toxic or oxidative damage to the blood–brain barrier itself increases its passive permeability, thus decreasing the efficiency of the protection afforded to the central nervous system.

Numerous new drugs have been recently developed for the therapy of brain disorders or diseases as a result of the identification of pharmacological targets by improvement of molecular biology technologies, receptor chemistry knowledge, and gene cloning. Frequently, however, the blood–brain barrier must be overcome to allow the administration of these neurotropic drugs. Whereas all organs of the body possess blood capillaries that allow an efficient exchange of solutes or drugs from the blood to the extracellular fluid, the CNS is perfused by capillaries formed by endothelial cells joined together by tight junctions of extremely low permeability that prevent any paracellular transport of ions, proteins, and other solutes (1); moreover, cerebral endothelial cells express ATP-dependent transmembrane glycoproteins, which reduce the intracellular accumulation of structurally and functionally unrelated xenobiotic compounds by extruding them out of cells, a process involved in multidrug resistance (2).

As a consequence, the BBB allows only the exchange by diffusion of lipophilic molecules following a concentration gradient, or by transcytosis mediated by transport systems located at the capillary endothelial cells. Because there are no fenestrations between the endothelial cells forming the BBB, lipophilic molecules present in the blood flow must first cross the luminal plasma membrane of endothelial cells, at a rate depending mainly on their liposolubility and molecular size, and on their eventual binding to plasma proteins. Since the brain exhibits complex distinct anatomical regions corresponding to functional specializations, such other factors as local blood perfusion intensity and microvessel density, developmental or aging state, and physiopathological events (inflammation, hypoxia) may also affect the rate of xenobiotic influx. After their entry into the endothelial cell, xenobiotics will encounter endothelial cytoplasmic enzymes potentially able to biotransform them (3). Since astrocyte endings are in close contact with the cerebral endothelial cells, glial enzymes could also take an active part in the metabolic protection of the brain.

For a long time, therefore, a hypothetical model of the BBB was that it functions like a lipid membrane. Recent progress on drug transport into or out of the brain, as well as a better knowledge of multidrug resistance, has demonstrated that this model is too simple.

II. PATHWAYS FOR DRUG ADMINISTRATION TO THE BRAIN

The major difficulty for cerebral drug administration lies in the existence of only three routes of administration that may be used for xenobiotic entry into the cerebral space (Fig. 1). These routes are through the BBB, the nasal route, and by intracerebroventricular (icv) administration or surgical implantation.

A. Through the Blood–Brain Barrier

For drugs administered by classical intravenous or intraperitoneal injections, or through the digestive tract, lung, or skin (i.e., drugs that are finally carried out by the bloodstream as in part 1 of Fig. 1), the route through the BBB allows relatively easy delivery. However, part of these lipophilic chemicals

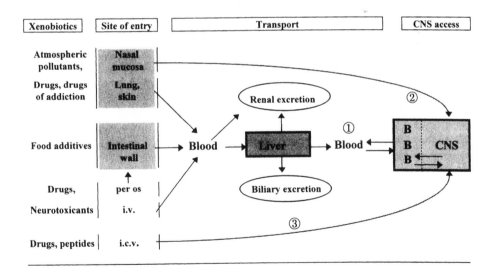

Fig. 1 Pathways for drug administration to the brain: 1, standard route (*i.e.*, through the BBB); 2, olfactory pathway; 3, surgical administration. Sites of drug metabolism are shaded.

is subject to being metabolized in the liver, resulting in modification in the amount of circulating drug available to the brain. Moreover, as depicted in parts 1 and 2 of Fig. 1, the liver, kidney, intestine, skin, lung, and also organs separating the brain from the blood flow, express enzymes able to metabolize xenobiotics. Another part of the dosage may be excreted by the kidney before entry into the CNS, rendering the precise amount of the drug that finally enters the brain difficult to estimate.

B. The Nasal Route

Chemicals like lipophilic atmospheric pollutants (solvents, engine smoke, aryl hydrocarbons), or drugs of addiction like cocaine or amphetamine derivatives, may rapidly enter the brain by the nasal route (part 2 of Fig. 1), and possibly promote reversible or irreversible alterations of nervous functioning (4). Anatomical pathways may allow molecular movement in the cerebrospinal fluid (CSF) from the subepithelial olfactory space and from the olfactory bulb, to the entorhinal cortex. Another possibility involves the transport in the axons of the olfactory neurons, as demonstrated for heavy metals, which can reach easily the brain in this way (5). Nasal drug intake has appeared to be a fast and effective route of administration, suitable for drugs that must act rapidly and are taken in small amounts, like antimigraine drugs, analgesics, vitamin B-12 or apomorphine. Nasal administration also prevents gastrointestinal clearance and hepatic metabolism and seems to be a promising route for the delivery of new pharmacologically active peptides to be produced by biotechnology (4).

It should be noted that if often used, this route of administration results frequently in attendant complications related to mucosal damage (e.g., infection, anosmia). Mammalian olfactory mucosa displays relatively high activity in the biotransformation of many xenobiotic compounds, catalyzed by active and preferentially or uniquely expressed enzyme isoforms (6, 7). This high metabolic activity should play an important role in protecting the underlying brain tissue from insults resulting from chronic environmental exposure to low doses of certain chemicals.

C. Intracerebroventricular Administration; Surgical Implantation

Part 3 of Fig. 1 describes drug delivery by icv devices (catheter, or osmotic pumps for intraventricular drug infusion) or by surgical implantation of devices that release an active molecule near its pharmacological target for variable time durations. Such methods allow the administration of a precise amount of drug to the brain; but they are invasive, hence restricted to a limited number of applications (e.g., the administration of water-soluble anticancer drugs, or treatment of intractable pain by direct central administration of morphine, preventing fatal side effects due to overdosage).

III. ENZYMES INVOLVED IN DRUG METABOLISM

Enzymes of drug metabolism are generally due to superfamilies of genes encoding an array of enzymes involved in the oxidation or conjugation of numerous lipophilic exogenous chemicals as well as of endogenous compounds, such as steroids, fatty acids, hormones, and prostaglandins (8). The main function of these enzymes, located mainly in the liver, was long considered to be to increase the water solubility of the parent drugs to improve their elimination in urine or bile. This does not represent an advantage for the brain, since polar metabolites will be entrapped inside the BBB-protected space, and unless specific transport systems for waste exist, will alter CNS homeostasis. The second function of these enzymes concerns the decrease or suppression of the pharmacological activity or of the toxicity of these compounds. However, some drug metabolites may be pharmacologically active: for instance, morphine 6-glucuronide is much more efficient as an analgesic than morphine itself (9). The intracerebral activation of lipophilic, inactive, but penetrating prodrugs to polar, efficient drugs presents useful applications in neuropharmacology, inasmuch as it is possible to design lipophilic drug precursors that will be metabolized in the brain to the active molecule (for a review, see Ref. 10). Unfortunately, xenobiotic biotransformation also often results in the formation of reactive metabolites that frequently bind to—or react with—proteins, lipids, or nucleic acids, leading to cytotoxicity and genotoxicity. Two important properties should be underlined:

1. A genetic polymorphism of several isoforms has been largely recognized and is frequently responsible for the biotransformation-dependent variability in drug response, bioavailability, and toxicity, as well as susceptibility to carcinogenesis or neurodegeneration (for recent reviews, see, e.g., Refs. 11–13).

2. Many genes coding for these enzymes are under complex and distinct control by both endogenous and exogenous compounds (8), allowing an adaptation of the activity to the amounts of certain substrates of these enzymes. Among endogenous compounds regulating gene expression, hormones, cytokines, and stress mediators have been studied extensively in the liver, but only

a few data concerning the regulation by sex hormones are available in the brain (14, 15). Concerning exogenous compounds, inducibility (i.e., increase of expression) of genes has been described after in vivo treatment with pheno-barbital, β-naphthoflavone (BNF), or 3-methylcholanthrene (3MC) for some isoforms in the brain (16–18), in isolated cerebral microvessels (3) and choroid plexus (19). Surprisingly, some cerebral cytochromes P450 (CYP) are induced by xenobiotics that are inefficient toward the corresponding hepatic isoforms. For example, administration of nicotine results in a selective CYP induction in some brain regions (20), but nicotine does not alter liver CYP expression. Moreover, the amplitude of the increase of expression of brain enzymes to an inducing treatment remains much lower than in the liver, suggesting different transcriptional regulations (21).

Drug-metabolizing enzymes have been classified in three phases, and their reactions are frequently sequential (Table 1).

A. Phase 1

1. *Monoamine oxidases A and B* (MAO-A and -B; EC 1.4.3.4) catalyze the oxidative deamination of numerous biogenic amines to the respective aldehyde and release of ammonia and hydrogen peroxide. These mitochondrial enzymes also represent a first line of protection for the brain, since they are expressed by cerebral microvessels and metabolize numerous neuroactive amines and their direct precursors, thus preventing disturbances of brain chemical message system by circulating molecules (22). The contribution of MAOs to xenobiotic metabolism has been now recognized (23). The high MAO-B activity in the endothelium of brain microvessels appear to efficiently protect the rat brain against neurotoxic exogenous pyridine derivatives (24) (see Sec. IV.A).

2. *Cytochromes P450* (CYP; EC 1.14.14.1) are heme proteins responsible both for the detoxification of drugs and other xenobiotics, which is achieved with very broad specificity, and for the metabolism of such endoge-

Table 1   Main Enzymatic Systems Involved in Drug Metabolism

Phase 1	Monoamine oxidases, flavoprotein-dependent monoxygenases, cytochromes P450, NADPH-cytochrome P450 reductase, epoxide hydrolases
Phase 2	UDP-glucuronosyltransferases, glutathione S-transferases, sulfotransferases, catechol O-methyltransferases
Phase 3	ATP-dependent transport of polar drugs and drug metabolites, catalyzed by Pgp and MRP

nous compounds as steroids and fatty acids, which is achieved with very high specificity. Cerebral CYP-dependent monoxygenase activities have been extensively studied because they can metabolize many lipophilic molecules able to cross the blood–brain barrier to more polar (hydroxylated) metabolites (16, 18, 21, 25–29). Reducing equivalents are supplied from NADPH to CYP by NADPH–cytochrome P450 reductase.

3. *Flavin-containing monoxygenases* (MFO; EC 1.14.13.8) consist in a group of NADPH– and oxygen-dependent oxygenases with broad specificity, able to oxygenate molecules having diverse functional groups containing phosphorus, nitrogen, sulfur, or selenium, as well as amines and hydrazines. Their activity does not require transfer of electrons by NADPH-cytochrome P450 reductase. Therefore MFOs work independently from the CYP–monoxygenase system. MFO-dependent activities have been demonstrated in the brain and in cultured astrocytes that metabolize N-methyl-4-phenyl-1,2,5,6-tetrahydropyridine (MPTP) to its N-oxide, a metabolite less toxic than the N-methylphenylpyridinium species, MPP$^+$ (30) (see Sec. IV.A). However, the most important part of MFO activity seems to be located in neurons (31).

4. *NADPH–cytochrome P450 reductase* (FP; EC 1.6.2.4) is a flavoprotein bound to the endoplasmic reticulum; its main function is to supply electrons from NADPH to CYP during the catalytic cycle of the latter. There is only one isoform identified in a given organism. The cerebral activity of this enzyme is important, since the ratio of CYP to FP is 10 times lower in the rat brain than in the liver (32, 33). If, therefore, the supply of electrons is the rate-limiting factor for CYP activity in the liver, in the brain, FP can independently reduce some substrates without affecting CYP activity. FP displays a high activity in isolated brain microvessels (3, 34) and cultured cerebrovascular endothelial cells (35).

5. *Epoxide hydrolases* (EH; EC 3.3.2.3), which are expressed as both soluble and microsomal isoforms, catalyze the conversion of a large variety of epoxides to less reactive *trans*-dihydrodiols. Microsomal epoxide hydrolase (mEH) displays a relatively low, but significant activity in the brain as well as in blood–brain interfaces (3, 28). In contrast to CYP, mEH mRNAs are expressed at stable amounts in the brain (36). mEH is involved in the metabolism of many xenobiotics of toxicological concern, including especially the metabolites of reactive epoxides formed from aryl hydrocarbons by CYP-catalyzed metabolism (37). Moreover, some large-sized metabolites like dihydroxybenzo[a]pyrene can be again metabolized by CYP to the powerful carcinogenic benzo[a]pyrene-7,8-diol-9,10-epoxide, which can no longer be hydrolyzed by EH for reasons of steric limitation (38, 39). Therefore, mEH could also play a role in activation processes.

B. Phase 2

1. *Uridine diphosphate glucuronosyltransferases* (UGT; EC 2.4.1.17) conjugate numerous hydroxylated substrates to glucuronic acid from UDP-glucuronic acid as a result of the expression of several isoforms that have distinct, but somewhat overlapping specificities (40). The first evidence of the glucuronidation of a drug (diazepam) in the primate brain was described more than 30 years ago (41). It is now well known that both rat and human brains possess relatively high 1-naphthol and 4-methylumbelliferone glucuronidation abilities (42, 43), suggesting that the brain expresses the UGT1A6 isoform. This finding has been confirmed recently by immunoblot assays using poly-clonal antibodies raised against the hepatic UGT1A6, and by reverse tran-scriptase polymerase chain reaction experiments showing the presence in the rat brain of a UGT1A6 mRNA identical to that found in the liver (44, 45). UGT1A6 has also been identified in cultured cerebral endothelial cells (35). However, the physiological substrate of this enzyme remains unidentified, since no significant glucuronidation of catecholamine neurotransmitters (nor-adrenalin, serotonin, dopamine) occurs in rat brain microsomes. The distribu-tion of 1-naphthol conjugation activity depends both on the region of the brain studied and on the type of cell, whereas sulfoconjugation remains low in brain regions possessing high glucuronoconjugation capacities. Concerning the cel-lular repartition, cultured astrocytes possess a twofold higher 1-naphthol con-jugation activity and a higher UGT1A6 expression than neurons (45). This result supports the old hypothesis that astrocytes should protect neurons from toxic insults. No immunoreactivity against the steroid reactive isoforms UGT2B2 and UGT2B3 (46), and no detectable activity toward morphine or bilirubin (42), were observed in the rat brain. On the other hand, there was a significant conjugation of paracetamol and naftazone, which are substrates of hepatic UGT1A6 (44). Interestingly, the expression of a few brain-specific isoforms (e.g., neolactotetraosylceramide glucuronyltransferase) that produce complex sulfate- and glucuronic acid–containing glycolipids, has been dem-onstrated (47, 48). These sulfoglucuronyl glycolipids play important roles in neuron and astrocyte adhesion and in cell–cell interactions during the develop-ment of the CNS.

2. *Glutathione S-transferases* (GST; EC 2.5.1.18) are a family of multifunctional dimeric enzymes that catalyze the conjugation of electrophilic xenobiotics with glutathione (GSH), which results in most cases in detoxifica-tion (49). GSTs are mainly located in the cytosol, but they are also active in mitochondria and in the endoplasmic reticulum. They also participate in the intracellular transport of a variety of hormones, endogenous metabolites, neu-

rotransmitters, and drugs, as a result of their capacity to bind these molecules. In the brain, GST activity is involved in the detoxification of the reactive quinone derivatives formed during catecholamine autoxidation, thus preventing redox cycling and subsequent oxidative stress (50) (see Sec. IV.C). Some isoforms of GST also catalyze the conjugation of 4-hydroxynonenal, a toxic product of lipoperoxidation, and therefore protect neurons from the consequences of an oxidative stress (51). On the other hand, GSTs are efficiently inhibited by MPTP metabolites (52), suggesting that GSTs could be one of the targets of MPTP neurotoxicity (see Sec. IV.A). Sex hormones play an important role in the regulation of brain GSTs (53), but the cofactor GSH does not appear to be under the influence of hormones.

3. *Phenol sulfotransferases* (PST; EC 2.8.2.1) are isoforms of a cytosolic multigenic enzyme family that catalyzes the sulfoconjugation of a variety of phenolic compounds of both endogenous or exogenous origin, such as aromatic monoamines, including catecholamine neurotransmitters, neurosteroids, and catecholamine metabolites (54). Therefore, PST and UGT frequently compete for the same substrate (e.g., 1-naphthol), resulting in a balance between their activities. The activities of PST isoforms are also regulated by separate genetic polymorphisms. PST isoforms are primarily located within neurons (55), and they control the levels of such physiologically active molecules as thyroid hormone, [Leu]enkephalin, and neurosteroids. Recent studies also show the implication of sulfation in the synthesis on neuronal cell surface of the extracellular matrix components chondroitin sulfate and keratan sulfate, proteoglycans being increasingly recognized as providing information relevant to cell–cell interactions and differentiation (56, 57). Drugs that are normally inactivated by sulfate conjugation may alter central endogenous metabolism by competition either for the same enzyme or for glutathione conjugation (58). Sulfoconjugation is a possible metabolic route for drugs such as salicylamide and methyldopa, and 1-naphthol is efficiently sulfoconjugated in rat brain cytosol and in the olfactory bulb (Suleman and Minn, unpublished results).

4. *Catechol O-methyltransferase* (COMT; EC: 2.1.1.6) is a ubiquitous enzyme metabolizing the endogenous catechols (e.g., catecholamines) involved in cerebral neurotransmission. Its localization is mainly glial, but COMT immunoreactivity has been detected in neurons (59). The dopamine precursor L-3,4-dihydroxyphenylalanine (levodopa) remains the most effective drug for the symptomatic treatment of Parkinson's disease, a degenerative neurological disorder related to a deficiency of the neurotransmitter dopamine. Being a catechol, levodopa is also a substrate for COMT; therefore the activity of the enzyme leads to the rapid decrease in the efficiency of the treatment. Recent improvements in therapy have made available several new antiparkinsonian drugs, which

are either dopamine agonists (among them, catechols like apomorphine) or COMT inhibitors, like tolcapone and entacapone, which are nitrocatechols (60).

Interfaces between the brain and its surroundings [i.e., brain capillaries (3), olfactory bulb (44), ependyma (61), and choroid plexus (19)] display 5- to 50-fold higher 1-naphthol conjugation activities than the brain cortex itself. Since some CYP isoforms, mEH, FP, and GST also display higher activities in the tissues of brain interface than in the brain (Table 2), we extended the concept of a *metabolic blood–brain barrier* [first postulated by Van Gelder (22), who described enzymatic activities toward amino acid neurotransmitters in isolated brain microvessels] to drug-metabolizing enzymes expressed in the cerebrovascular endothelial cells. These enzymes protect the brain, at least partially, against lipophilic endogenous molecules or xenobiotics able to cross the *physical blood–brain barrier* (21).

These activities change with brain development and aging. Especially in the rat olfactory bulb, high glucuronidation activities toward 1-naphthol and odorant molecules are expressed during the first weeks of perinatal life, corresponding probably to the need of access to food, the visual system being not efficient during the first days of postnatal life. These high activities observed in the neonatal olfactory bulb decrease to stable adult (3-month-old rat) values during brain maturation (44).

C. Phase 3

Since the metabolites formed by phases 1 and 2 drug metabolism are polar, their hydrosolubility prevents their elimination from the cell by diffusion

Table 2 Ratio of Drug Metabolism Activities in Isolated Brain Microvessel, Olfactory Bulb, and Choroid Plexus Homogenates to Brain Cortex Homogenate

Enzyme	Brain microvessels[a]	Olfactory bulb[a]	Choroid plexus[a]
CYP2B	1.0	3.2	ND
CYP1A + CYP2B	3.9	4.2	20.0
NADPH–CYP reductase	1.3	ND	2.4
UGT1A6[b]	1.5	5.2	>50
mEH	5.4	0.9	36.0
GST[c]	1.8	1.8	2.1
PST[b]	ND	5.2	1.0

[a] ND, not determined.
[b] Activity measured using 1-naphthol as a substrate.
[c] Activity measured using *trans*-stilbene oxide (TSO) as a substrate.

through the plasma membrane. As a result, xenobiotic metabolites may be entrapped inside the CNS space by the blood–brain barrier, altering cerebral extracellular homeostasis. They may also endanger neuron functioning, especially if pharmacologically active or reactive metabolites are formed. Therefore, it was interesting to study the mechanisms potentially involved in the elimination of intracerebrally formed metabolites and to determine which belong to *phase 3* of drug metabolism. Previous in vitro and in vivo studies carried out in our group suggested the presence of a specific active transporter allowing a rapid clearance of 1-naphthyl glucuronide from the brain (62). On the other hand, it has been shown that chemotherapy against brain cancers is often ineffective because brain endothelial cells possess mechanisms that enable them to prevent the influx of the cytotoxic agent. This multidrug resistance (MDR) results from the high expression in endothelial cells of Pgp, a 170 kDa membrane glycoprotein that appears to be a member of the ATP-binding cassette family of transporters (63), and also from the activity of multidrug resistance–associated protein (MRP), a 190 kDa protein structurally related to Pgp. Both ATP-dependent transporters can exclude numerous cytotoxic drugs, but also probably glutathione conjugates or glucuronides (64). Moreover, immunoreactivity to Pgp has been recently found in astrocytes (65), suggesting that astrocytes participate to the efflux system at the blood–brain barrier. This remains to be clearly demonstrated at blood–brain interfaces.

IV. SIDE EFFECTS OF DRUG METABOLISM

The beneficial effects of cerebral drug metabolism are obvious: the decrease or suppression of pharmacological effects of neuroactive drugs allows a suitable control of nervous activity; the decrease or suppression of xenobiotic toxicity prevents neuronal dysfunctioning or death; and phase 3 systems actively exclude both drugs and their metabolites from the cerebral extracellular space, thus allowing the maintenance of adequate brain homeostasis. Nevertheless, this ideal picture is much too optimistic, since undesirable effects of drug metabolism, endangering normal functioning of both blood–brain barrier and brain, are frequently observed.

A. Formation of Reactive Metabolites

Drug metabolites are not always safe. Xenobiotic metabolism frequently produces reactive metabolites that could directly or indirectly alter brain function and neuron survival.

1. Monoamine Oxidase Inhibitors

The metabolism of neurotransmitter amines by cerebral MAO-A and -B produces equivalent amounts of the corresponding aldehydes, and ammonia. Both products are considered to be neurotoxic and could participate in the progressive neuronal death that occurs during senescence. Moreover, implication of MAOs in xenobiotic metabolism was definitely assessed by the demonstration that MPTP, a meperidine analogue promoting a parkinsonian-like syndrome in humans (66), monkeys (67), and mice (68), was a substrate of MAO-B in most mammalian species assayed (69). The precise mechanism by which MPTP administration induces selective dopaminergic neuronal death in the pars compacta of the substantia nigra is still a subject of debate. Exposure to MPTP results in the formation of MPP^+, which is actively taken up by dopaminergic neurons and kills them either by blocking complex 1 of the mitochondrial respiratory chain or by promoting the formation of superoxide and hydroxyl radicals (for a recent review, see Ref. 70). Defects in both mitochondrial complex 1 and oxidative stress are considered to be the pathological processes that may cause the degeneration of dopaminergic neurons in idiopathic Parkinson's disease (71). MPP^+ is a polar compound: it does not cross the BBB, and accordingly, an intravenous MPP^+ injection does not promote neurotoxicity.

MPTP has emerged as an environmental molecule allowing the design of animal models of Parkinson's disease, which is a specific human neurodegeneration of striatal dopaminergic neurons, the etiology of which remains largely unknown (70). Investigations of MPTP also allowed the finding of endogenous molecules that are structurally related to MPP^+, such as isoquinoline and β-carboline derivatives. These molecules are oxidized by MAO to reactive isoquinolinium or β-carbolinium species that inhibit mitochondrial respiration and promote oxidative stress. A deregulation of their normal metabolism could participate in the etiology of Parkinson's disease (for a recent review, see Ref. 72). These results suggest that administration of L-deprenyl, a specific MAO-B inhibitor, is of great potential in the treatment of patients with Parkinson's disease.

The very high MAO-B activities observed in isolated rat brain microvessels have been supposed to protect the brain against MPTP toxicity (24). This seems to be confirmed by a report that an intranigral infusion of low quantities of MPP^+ destroyed rat dopaminergic neurons (74).

2. Flavin-Containing Monoxygenases and Cytochromes P450

NADPH-dependent monoxygenases may easily form diverse potentially harmful metabolites (e.g., epoxides, nitrosamines, sulfoxides, imminium species, aldehydes) that may form protein adducts and cytosolic antigens, or promote more or less reversible neurotoxicity. In this chapter, for evident space and topic reasons, we review only some examples of neurotoxicity resulting from cerebral metabolism of drugs or xenobiotics.

A first example concerns the metabolism of the phosphorothionate insecticide parathion to a potent cerebral anticholinesterase metabolite, paraoxon (75). The parent molecule is activated in vivo by CYP-mediated desulfuration in close vicinity of the target enzyme. The environmental use of that molecule may therefore result in severe neurotoxicity.

Nicotine appears to be both a psychostimulant drug and a common environmental pollutant promoting ''passive smoking.'' It should be also considered to be a drug of addiction, since thousands of heavy tobacco smokers are unable to give up smoking, showing therefore a true dependence on nicotine. Smoking and inhalation are routes of administration that allow a very rapid delivery of the drug to the brain, and the first daily puff on a cigarette is considered by tobacco smokers to be ''the best one'' because it efficiently attenuates the morning's withdrawal symptoms. Nicotine is a pharmacologically active tertiary amine, efficiently metabolized to cotinine by both liver and brain CYP (76) through the formation of a reactive nicotine-$\Delta[1'(5')]$-imminium ion, which is an alkylating species (77). During its metabolism, nicotine undergoes covalent binding to microsomal protein, supporting the concept that reactive metabolic intermediates may play a role in the pharmacology and toxicity of nicotine.

Phencyclidine, originally developed as an anesthetic and analgesic in dentistry, was withdrawn from human use because of significant side effects, including agitation, blurred vision and hallucinations, and paranoid behavior (78). Owing to its hallucinogenic properties, PCP has become a popular drug of abuse, with different behavioral responses among individuals. Liver biotransformation of PCP results in the formation of numerous mono- or dihydroxylated metabolites, suggesting the involvement of several CYP isoforms in this metabolic pathway (79) and explaining in part the different behavioral responses to the drug resulting from genetic polymorphism among individuals (80). PCP is also metabolized in the brain, forming both inactive and pharmacologically active metabolites near their receptors (81, 82). Moreover, PCP metabolism promotes its covalent binding to macromolecules and inactivation of hepatic CYP2B1 through the NADPH-dependent α-carbon oxidation of PCP, which leads to the formation of the electrophilic PCP imminium ion (83). This irreversible inhibitory effect has also been demonstrated in the rat brain, where PCP inactivates specifically CYP-catalyzed ethoxyresorufin dealkylation (Perrin and Minn, unpublished results).

The selective irreversible binding of reactive intermediates formed by CYP1A metabolism has also been used as a marker of metabolism in some tissues or cells. A high binding in the endothelial cells of the capillary loops of the choroid plexus was described in 1994, suggesting an efficient, BNF-inducible, CYP1A-dependent metabolism in these endothelial cells (84).

3. Epoxide Hydrolases

No experimental data concerning cerebral mEH-promoted toxicity is available. The anticonvulsants phenytoin and carbamazepine, which possess aryl moieties, are metabolized in the human liver to epoxide intermediates responsible for hepatic necrosis (85). To our knowledge, however, no evidence of such deleterious effects either at the blood–brain barrier or to the brain has been presented. It should be of importance that the human brain displays a 40-fold higher mEH activity than the rat brain (43).

4. Conjugation Enzymes

Enzymes catalyzing phase 2 of drug metabolism may also form reactive metabolites. For instance, UGT may promote the formation of protein adducts when it is conjugating carboxylic acids like nonsteroidal anti-inflammatory drugs (NSAIDs) (86). But to our knowledge, even if there is evidence that some NSAIDs cross the blood–brain barrier and enter the CNS, the possible formation of reactive metabolites in the brain has never been reported. Formation of reactive metabolites has been also described during hepatic conjugation of arylamines or polycyclic arylmethanol with sulfate and glutathione; but once again, no evidence of such activities in the brain has been presented.

The neurotoxicity of a number of molecules has been already well documented, but it seems probable that the long-term intake or consumption of rather common pesticides, herbicides, drug additives, or more or less illicit drugs will be related in the relatively near future to neurological disorders or neurodegenerative diseases.

B. Oxygen Reduction Products

The potential toxicity of oxygen-related side products formed during drug metabolism activities remains a fascinating topic. It has been clearly established that oxygen radicals can peroxidize lipids, thus modifying many membrane properties such as fluidity; these radicals also can depolymerize proteoglycans (thus altering cell–cell adhesion and communication), oxidize proteins (resulting in changes in enzyme activities, receptor affinities, transport kinetics, and immunological responsiveness), and finally, alter nucleic acids and their genomic content. In fact, all cell macromolecules are potential targets that can be altered by oxidative stress, resulting frequently in aging-like alterations of their properties (87). Therefore, the reactivity of oxygen radicals appears to be of paramount importance because of their ability to modify many molecular characteristics and specific properties allowing the functioning of both the BBB (88) and the CNS (89). The brain continuously produces superoxide radicals as by-products of aerobic metabolism, and this lifelong radical production results in the formation of abnormal mitochondria and lipofuscin deposits that are characteristic markers of brain aging. The brain is also the most sensitive tissue to oxidative damage, owing to its high and constant requirement for oxygen, as well as high amounts of easily oxidizable or peroxidizable substrates, and non-protein-bound iron in the cerebrospinal fluid.

The formation of oxygen reduction products (superoxide, hydrogen peroxide, or hydroxyl radicals) has been described during the activities of several phase 1 enzymatic systems. The superoxide anion is not considered to be very toxic by itself, but it may form either harmful hydroxyl radicals by Fenton kinetics in the presence of transition metals (90) or peroxynitrite in the presence of nitric oxide (91). The latter is considered to be responsible for most pathological oxidative stress in the living tissue (for a review, see Ref. 92). Since astrocytes possess more efficient scavenging systems than neurons (93), oxidative stress may result in selective neuronal death by promoting necrosis or apoptosis, and an associated increase in passive permeability of the BBB may contribute to human neurodegenerative pathologies such as the following.

1. MAO activity involves the formation of hydrogen peroxide, and this enzyme has been suspected to participate in the global increase of oxidized products in the aged brain (94) and in oxidative damage to the substantia nigra observed in Parkinson's disease (95). Therefore, deprenyl, a specific MAO-B inhibitor, and also a molecule protecting dopamine neurons from peroxynitrite-promoted apoptosis (96), is frequently coadministered with dopamine agonists for the treatment of Parkinson's disease (73).

2. CYP-dependent activities continuously generate superoxide, as a "leak" of reduced oxygen during the catalytic cycle (97). This radical production is increased by inducers of the expression of some CYP isoforms (3MC or BNF) and inhibited by the depletion of NADPH or oxygen, or by selective inhibition of CYP activities. Because superoxide radical may promote the formation of other more toxic radicals, the limited capacity of the cerebral en-

zymes to be induced by 3MC or BNF can be considered to be a protection mechanism against high activity-promoted oxidative stress.

3. Molecular radicals formed by the one-electron reduction of some xenobiotics by NADPH-cytochrome P450 reductase may react with molecular oxygen to form superoxide. Many quinones display cytotoxic properties that have made them useful as anticancer or antibacterial drugs. The molecular mechanism of their cytotoxicity is generally considered as an enzymatic redox cycling with the participation of NADH- or NADPH-dependent reductases. The enzymatic reduction of quinones results in the formation of semiquinones, followed in aerobic conditions by a reduction of oxygen to superoxide and the regeneration of the parent quinone (98) (Fig. 2). A significant superoxide production during xenobiotic metabolism has been demonstrated in isolated brain microvessels or choroid plexus homogenates and in brain microsomes (34, 99).

The microsomal enzyme involved in superoxide formation by one electron-step reduction of quinones and nitroheterocyclic and imminium compounds has been identified as NADPH-cytochrome P450 reductase. Xenobiotic-promoted superoxide production has also been observed with neurons, astrocytes, and cerebrovascular endothelial cells in primary cultures, the rate of molecule entry into the cell being a reaction-limiting step (35, 100). Since cerebrovascular endothelial cells are continuously exposed to blood-borne molecules, their metabolic ability toward some xenobiotics may result in the formation of reactive metabolites or oxygen species, probably altering their specific properties, especially their selective permeability (101–103) and also

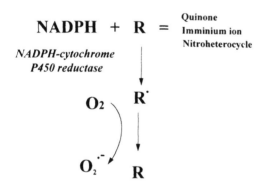

Fig. 2 Formation of superoxide radicals during the reductive metabolism of xenobiotics by NADPH–cytochrome P450 reductase in normoxic conditions (redox cycling).

resulting in a dose-dependent increase in lipid peroxidation products (unpublished observations).

C. Nonenzymatic Oxidation of Drugs

The nonenzymatic autoxidation of catecholamines also plays an important role in the physiology and aging of the CNS. Dopamine, like most catecholamines, can be easily oxidized by molecular oxygen in physiological solutions (i.e., at neutral pH and in the presence of transition metal traces) (104). During autoxidation, both semiquinones and quinones are formed, and they react with molecular oxygen to produce reactive oxygen species. Numerous data suggest that the cytotoxicity of levodopa, a dopamine precursor used for long-term therapy of Parkinson's disease, is likely due to the action of free radicals formed as a result of its autoxidation (105, 106). Moreover, quinoid compounds derived from the autoxidation of endogenous catechols polymerize to form neuromelanin, which contributes to the vulnerability of dopaminergic neurons in Parkinson's disease (107). Finally, quinoids products can cross-link with neurofilament proteins (108) and with cysteine to form cytotoxic cysteinylcatechols (109). This process may contribute to the formation of Lewy bodies, neurofilamentous cytoplasmic inclusions closely linked with nigral neurodegeneration in Parkinson's disease (110). Nevertheless, some contradictory results indicate that both apomorphine and levodopa may act as antioxidants, depending in fact on their concentration. At low concentrations, these catechols act as efficient scavengers of reactive oxygen species, efficiently protecting cultured neurons against promoters of lipid peroxidation; but at higher levels, they promote a cytotoxic oxidative stress (111, 112). Moreover, the up-regulation of cellular GSH evoked by autoxidizable agents is associated with significant protection of mixed cultures (neurons plus glia), suggesting that an ability to up-regulate GSH may serve a protective role in vivo (113, 114).

There is now a wide variety of drugs and formulations available for the treatment of Parkinson's disease. These include anticholinergics, amantidine, levodopa, apomorphine and new dopamine agonists, selegiline, and more recently COMT inhibitors (115, 116). COMT is responsible for the degradation of levodopa and dopamine; therefore, inhibiting COMT activity is one method of extending the action of these molecules. The new nitrocatechol-type COMT inhibitors entacapone, nitecapone, and tolcapone inhibit COMT in the periphery; tolcapone also inhibits COMT activity centrally (117). Since several catechols (e.g., levodopa and apomorphine) that are currently used as dopamine agonists display some in vitro cytotoxicity (112, 118) and are suspected of accelerating the progression of the disease (105), eventual possible side effects of nitrocatechols need to be precisely studied.

V. CONCLUSIONS

A large body of experimental data largely suggests that drug-metabolizing activities, and especially conjugation, participate in CNS defense against toxic assaults both at brain interfaces and in the cerebral parenchyma itself. These enzymatic systems also participate in more physiological functions such as the metabolism of endogenous neuroactive substances and the maintenance of homeostasis in the cerebral extracellular fluid. In contrast, among these enzymatic systems, several may produce during the metabolism of xenobiotics either reactive metabolites or free radicals, able to alter macromolecular components of both neurons and cerebrovascular endothelial cells. Therefore, chronic administration of some neurotropic drugs may promote blood–brain barrier dysfunction and possibly accelerate oxidative mechanisms involved in brain aging. Most of the data presented in this chapter result from in vitro or in vivo animal experiments. Humans exhibit important interindividual differences in drug sensitivity and metabolic capacity, mainly resulting from genetic polymorphism. This polymorphism might contribute to the effects of pollutants and other environmental chemicals on susceptibility to cancer and Parkinson's disease, suggesting a role of environmental toxins in the etiology of neurodegenerative disorders. The preferential distribution of phase 1 enzymes in neurons should also be an important factor in neuronal damage promoted by drug metabolism. On the other hand, if free radical production presents evident pathological aspects, it could also be related to specific physiological functions. Recent studies provide evidence for a role of secondary messenger for some free radicals in regulation of cell signaling, gene expression, receptor or transporter affinities, immunological responses, and enzymatic activities (see, e.g., Refs. 119–121). The mechanism of such redox regulations remains unclear, especially concerning brain function, but probably will be of interest as a new drug target in a few years.

ACKNOWLEDGMENTS

D.G. and R.S.E.B. were research fellows of the Sandoz Foundation for Gerontological Research (Basel, Switzerland) and the Fundação Coordenação de

Aperfeiçoamento de Pessoal de Nível Superior (Brazil), respectively. This work was made possible by the financial help of the Fondation pour la Recherche Médicale and from BIOMED BMH-97-2621. We are grateful to Drs. A. G. de Boer and W. Sutanto (Leiden, Netherlands) for organizing workshops under the program EC BIOMED Concerted Action "Drug Transport to the Brain: New Experimental Strategies," and to Dr. J. Magdalou for critical reading of the manuscript.

REFERENCES

1. LL Rubin. Endothelial cells: Adhesion and tight junctions. Curr Opin Cell Biol 4:830–833, 1992.
2. J Van Asperen, U Mayer, O van Tellingen, JH Beijnen. The functional role of P-glycoprotein in the blood–brain barrier. J Pharm Sci 86:881–884, 1997.
3. JF Ghersi-Egea, A Minn, G Siest. A new aspect of the protective functions of the blood–brain barrier: Activities of four drug-metabolizing enzymes in isolated brain microvessels. Life Sci 42:2515–2523, 1988.
4. WM Pardridge. Transnasal and intraventricular delivery of drugs. In: WM Pardridge, ed. Peptide Drug Delivery to the Brain. New York: Raven Press, 1991, pp 99–122.
5. L Hastings, JE Evans. Olfactory primary neurons as a route of entry for toxic agents into the CNS. Neurotoxicology 12:707–714, 1991.
6. D Lazard, N Tal, M Rubinstein, M Khen, D Lancet, K Zupko. Identification and biochemical analysis of novel olfactory-specific cytochrome P-450IIA and UDP-glucuronosyl transferase. Biochemistry 29:7433–7440, 1990.
7. CJ Reed. Drug metabolism in the nasal cavity: Relevance to toxicology. Drug Metab Rev 25:173–205, 1993.
8. DW Nebert. Proposed role of drug-metabolizing enzymes: Regulation of steady state levels of the ligands that effect growth, homeostasis, differentiation, and neuroendocrine functions. Mol Endocrinol 5:1203–1214, 1991.
9. FV Abbott, RM Palmour. Morphine-6-glucuronide: Analgesic effects and receptor binding profile in rats. Life Sci 43:1685–1695, 1988.
10. N Bodor, P Buchwald. Drug targeting via retrometabolic approaches. Pharmacol Ther 76:1–27, 1997.
11. RC Strange, JT Lear, AA Fryer. Glutathione S-transferase polymorphisms: Influence on susceptibility to cancer. Chem Biol Interact 111–112:351–364, 1998.
12. T Yokoi, T Kamataki. Genetic polymorphism of drug metabolizing enzymes: New mutations in CYP2D6 and CYP2A6 genes in Japanese. Pharm Res 15: 517–524, 1998.

13. AG Riedl, PM Watts, P Jenner, CD Marsden. P450 enzymes and Parkinson's disease: The story so far. Mov Disord 13:212–220, 1998.

14. R Perrin, A Minn, JF Ghersi-Egea, MC Grassiot, G Siest. Distribution of cytochrome P-450 activities towards alkoxyresorufin derivatives in rat brain regions, subcellular fractions and cerebral microvessels. Biochem Pharmacol 40: 2145–2151, 1990.

15. HK Anandatheerthavarada, V Ravindranath. Administration of testosterone alleviates the constitutive sex difference in rat brain cytochrome P-450. Neurosci Lett 125:238–240, 1991.

16. HW Strobel, E Cattaneo, M Adesnik, A Maggi. Brain cytochromes P-450 are responsive to phenobarbital and tricyclic amines. Pharmacol Res 21:169–175, 1989.

17. A Dhawan, D Parmar, M Das, PK Seth. Cytochrome P-450 dependent monooxygenases in neuronal and glial cells: Inducibility and specificity. Biochem Biophys Res Commun 170:441–447, 1990.

18. SV Bhagwat, MR Boyd, V Ravindranath. Brain mitochondrial cytochromes P450: Xenobiotic metabolism, presence of multiple forms and their selective inducibility. Arch Biochem Biophys 320:73–83, 1995.

19. B Leininger-Muller, JF Ghersi-Egea, G Siest, A Minn. Induction and immunological characterization of the uridine diphosphate–glucuronosyltransferase conjugating 1-naphthol in the rat choroid plexus. Neurosci Lett 175:37–40, 1994.

20. HK Anandatheerthavarada, JF Williams, L Wecker. Differential effect of chronic nicotine administration on brain cytochrome P4501A1/2 and P4502E1. Biochem Biophys Res Commun 194:312–318, 1993.

21. A Minn, JF Ghersi-Egea, R Perrin, B Leininger, G Siest. Drug metabolizing enzymes in the rat brain and brain microvessels. Brain Res Rev 16:65–82, 1991.

22. NM Van Gelder. A possible enzyme barrier for gamma-aminobutyric acid in the central nervous system. Prog Brain Res 29:259–271, 1968.

23. M Strolin-Benedetti, P Dostert. Contribution of monoamine oxidases to the metabolism of xenobiotics. Drug Met Rev 26:507–535, 1994.

24. NJ Riachi, SI Harik. Strain differences in systemic 1-methyl-4-phenyl-1,2,3,6-tetrahydropyridine neurotoxicity in mice correlate best with monoamine oxidase activity at the blood–brain barrier. Life Sci 42:2359–2363, 1988.

25. M Das, H Mukhtar, PK Seth. Aryl hydrocarbon hydroxylase and glutathione-S-transferase activities in discrete regions of rat brain. Toxicol Lett 13:125–128, 1982.

26. B Walther, JF Ghersi-Egea, Z Jayyosi, A Minn, G Siest. Ethoxyresorufin O-deethylase activity in rat brain subcellular fractions. Neurosci Lett 76:58–62, 1987.

27. BM Naslund, H Glaumann, M Warner, JA Gustafsson, T Hansson. Cytochrome P-450 b and c in the rat brain and pituitary gland. Mol Pharmacol 33:31–37, 1988.

28. JF Ghersi-Egea, B Leininger-Muller, G Suleman, G Siest, A Minn. Localization of several drug-metabolizing enzymes activities in blood–brain interface structures. J Neurochem 62:1089–1096, 1994.

29. V Ravindranath, MR Boyd. Xenobiotic metabolism in brain. Drug Metab Rev 27:419–448, 1995.

30. DA Di Monte, EY Wu, I Irwin, LE Delanney, JW Langston. Biotransformation of 1-methyl-4-phenyl-1,2,3,6-tetrahydropyridine in primary cultures of mouse astrocytes. J Pharmacol Exp Ther 258:594–600, 1991.

31. S Bhamre, SV Bhagwat, SK Shankar, DE Williams, V Ravindranath. Cerebral flavin-containing monooxygenase-mediated metabolism of antidepressants in brain: Immunochemical properties and immunocytochemical localization. J Pharmacol Exp Ther 267:555–559, 1993.

32. JF Ghersi-Egea, A Minn, JL Daval, Z Jayyosi, V Arnould, H Souhaili-El Amri, G Siest. NADPH:cytochrome P-450(c) reductase: Biochemical characterization in rat brain and cultured neurones and evolution of activity during development. Neurochem Res 14:883–888, 1989.

33. V Ravindranath, HK Anandatheerthavarada, SK Shankar. NADPH cytochrome P450 reductase in rat, mouse and human brain. Biochem Pharmacol 39:1013–1018, 1990.

34. P Lagrange, MH Livertoux, MC Grassiot, A Minn. Superoxide anion formation during monoelectronic reduction of xenobiotics by preparations of rat brain cortex, microvessels and choroid plexus. Free Radical Biol Med 17:355–359, 1994.

35. M Chat, C Bayol-Denizot, G Suleman, F Roux, A Minn. Drug metabolizing enzyme activities and superoxide formation in primary and immortalized rat brain endothelial cells. Life Sci 62:151–163, 1998.

36. B Schilter, CJ Omiecinski. Regional distribution and expression modulation of cytochrome P-450 and epoxide hydrolase mRNAs in the rat brain. Mol Pharmacol 44:990–996, 1993.

37. AY Lu, GT Miwa. Molecular properties and biological functions of microsomal epoxide hydrase. Annu Rev Pharmacol Toxicol 20: 513–531, 1980.

38. P Bentley, F Oesch, H Glatt. Dual role of epoxide hydratase in both activation and inactivation of benzo(a)pyrene. Arch Toxicol 39:65–75, 1977.

39. E Huberman, L Sachs. DNA binding and its relationship to carcinogenesis by different polycyclic hydrocarbons. Int J Cancer 19:122–127, 1977.

40. PI Mackenzie, IS Owens, B Burchell, KW Bock, A Bairoch, A Belanger, S Fournel-Gigleux, M Green, DW Hum, T Iyanagi, D Lancet, P Louisot, J Magdalou, JR Chowdhury, JK Ritter, H Schachter, TR Tephly, KF Tipton, DW Nebert. The UDP glycosyltransferase gene superfamily: Recommended nomenclature update based on evolutionary divergence. Pharmacogenetics 7:255–269, 1997.

41. G Benzi, F Berte, A Crema, GM Frigo. Cerebral drug metabolism investigated by isolated perfused brain in situ. J Pharm Sci 56:1349–1351, 1967.

42. FG Suleman, B Leininger-Muller, JF Ghersi-Egea, A Minn. Uridine diphos-

phate–glucuronosyltransferase activities in rat brain microsomes. Neurosci Lett 161:219–222, 1993.

43. JF Ghersi-Egea, R Perrin, B Leininger-Muller, MC Grassiot, C Jeandel, J Floquet, G Cuny, G Siest, A Minn. Subcellular localization of cytochrome P450, and activities of several enzymes responsible for drug metabolism in the human brain. Biochem Pharmacol 45:647–658, 1993.

44. D Gradinaru, FG Suleman, J Magdalou, A Minn. UDP-glucuronosyltransferase in the rat olfactory bulb: Identification of the UGT1A6 isoform and age-related changes in 1-naphthol glucuronidation. Neurochem Res 24:995–1000, 1999.

45. FG Suleman, A Abid, D Gradinaru, JL Daval, J Magdalou, A Minn. Characterization of a uridine diphosphate glucuronosyltransferase isoform similar to rat liver UGT1A6 in rat brain and cerebral cells in cultures. Arch Biochem Biophys 358:63–67, 1998.

46. MK Martinasevic, CD King, GR Rios, TR Tephly. Immunohistochemical localization of UDP-glucuronosyltransferases in rat brain during early development. Drug Metab Dispos 26:1039–1041, 1998.

47. DK Chou, S Flores, FB Jungalwala. Expression and regulation of UDP-glucuronate: Neolactotetraosylceramide glucuronyltransferase in the nervous system. J Biol Chem 266:17941–17947, 1991.

48. K Terayama, S Oka, T Seiki, Y Miki, A Nakamura, Y Kozutsumi, K Takio, T Kawasaki. Cloning and functional expression of a novel glucuronyltransferase involved in the biosynthesis of the carbohydrate epitope HNK-1. Proc Natl Acad Sci USA 94:6093–6098, 1997.

49. M Abramowitz, H Homma, S Ishigaki, F Tansey, W Cammer, I Listowsky. Characterization and localization of glutathione S-transferases in rat brain and binding of hormones, neurotransmitters, and drugs. J Neurochem 50:50–57, 1988.

50. S Baez, J Segura-Aguilar, M Widersten, AS Johansson, B Mannervik. Glutathione transferases catalyse the detoxication of oxidized metabolites (o-quinones) of catecholamines and may serve as an antioxidant system preventing degenerative cellular processes. Biochem J 324:25–28, 1997.

51. SS Singhal, P Zimniak, S Awasthi, JT Piper, NG He, JI Teng, DR Petersen, YC Awasthi. Several closely related glutathione S-transferase isozymes catalyzing conjugation of 4-hydroxynonenal are differentially expressed in human tissues. Arch Biochem Biophys 311:242–250, 1994.

52. YC Awashi, VS Shivendra, R Sing, GW Abell, W Gessner, Brossi A. MPTP metabolites inhibit rat brain glutathione S-transferases. Neurosci Lett 81:159–163, 1987.

53. M Das, AK Agarwal, PK Seth. Regulation of brain and hepatic glutathione-S-transferase by sex hormones in rats. Biochem Pharmacol 31:3927–3930, 1982.

54. AJ Rivett, BJ Eddy, JA Roth. Contribution of sulfate conjugation, deamination, and O-methylation to metabolism of dopamine and norepinephrine in human brain. J Neurochem 39:1009–1016, 1982.

55. JA Roth. Sulfoconjugation: Role in neurotransmitter and secretory protein activity. Trends Pharmacol Sci 7:404–407, 1986.

56. H Kitagawa, K Tsutsumi, Y Tone, K Sugahara. Developmental regulation of the sulfation profile of chondroitin sulfate chains in the chicken embryo brain. J Biol Chem 272:31377–31381, 1997.

57. C Lander, H Zhang, S Hockfield. Neurons produce a neuronal cell surface–associated chondroitin sulfate proteoglycan. J Neurosci 18:174–183, 1998.

58. A Baranczyk-Kuzma, J Sawicki. Biotransformation in monkey brain: Coupling of sulfation to glutathione conjugation. Life Sci 61:1829–1841, 1997.

59. T Karhunen, C Tilgmann, I Ulmanen, P Panula. Neuronal and non-neuronal catechol-O-methyltransferase in primary cultures of rat brain cells. Int J Dev Neurosci 13:825–834, 1995.

60. MD Gottwald, JL Bainbridge, GA Dowling, MJ Aminoff, BK Alldredge. New pharmacotherapy for Parkinson's disease. Ann Pharmacother 31:1205–1217, 1997.

61. MR Del Bigio. The ependyma: A protective barrier between brain and cerebrospinal fluid. Glia 14:1–13, 1995.

62. B Leininger, JF Ghersi-Egea, G Siest, A Minn. In vivo study of the elimination from brain tissue of an intracerebrally-formed xenobiotic metabolite, 1-naphthyl-β-D-glucuronide. J Neurochem 56:1163–1168, 1991.

63. A Regina, A Koman, M Piciotti, B El Hafny, MS Center, R Bergmann, PO Couraud, F Roux. Mrp1 multidrug resistance–associated protein and P-glycoprotein expression in rat brain microvessel endothelial cells. J Neurochem 71:705–715, 1998.

64. D Keppler, I Leier, G Jedlitsky. Transport of glutathione conjugates and glucuronides by the multidrug resistance proteins MRP1 and MRP2. Biol Chem 378:787–791, 1997.

65. WM Pardridge, PL Golden, YS Kang, U Bickel. Brain microvascular and astrocyte localization of P-glycoprotein. J Neurochem 68:1278–1285, 1997.

66. JW Langston, P Ballard, JW Tetrud, I Irwin. Chronic Parkinsonism in humans due to a product of meperidine-analog synthesis. Science 219:979–980, 1983.

67. W Schultz, A Studer, G Jonsson, E Sundström, I Mefford. Deficits in behavioral initiation and execution processes in monkeys with 1-methyl-4-phenyl-1,2,3,6-tetrahydropyridine-induced parkinsonism. Neurosci Lett 59:225–232, 1985.

68. RE Heikkila, PK Sonsalla. The MPTP-treated mouse as a model of parkinsonism: How good is it? Neurochem Int 20:299S–303S, 1992.

69. K Chiba, A Trevor, N Castagnoli Jr. Metabolism of the neurotoxic tertiary amine, MPTP, by brain monoamine oxidase. Biochem Biophys Res Commun 120:547–578, 1984.

70. E Bezard, C Imbert, CE Gross. Experimental models of Parkinson's disease: From the static to the dynamic. Rev Neurosci 9:71–90, 1998.

71. MF Beal. Aging, energy, and oxidative stress in neurodegenerative diseases. Ann Neurol 38:357–366, 1995.

72. KS McNaught, PA Carrupt, C Altomare, S Cellamare, A Carotti, B Testa, P Jenner, CD Marsden. Isoquinoline derivatives as endogenous neurotoxins in the aetiology of Parkinson's disease. Biochem Pharmacol 56:921–933, 1998.

73. CW Olanow, RA Hauser, L Gauger, T Malapira, W Koller, J Hubble, K Bushenbark, D Lilienfeld, J Esterlitz. The effect of deprenyl and levodopa on the progression of Parkinson's disease. Ann Neurol 38:771–777, 1995.

74. LM Sayre, Arora PK, LA Iacofano, SI Harik. Comparative toxicity of MPTP, MPP$^+$ and 3,3-dimethyl-MPDP$^+$ to dopaminergic neurons of the rat substantia nigra. Eur J Pharmacol 124:171–174, 1986.

75. CS Forsyth, JE Chambers. Activation and degradation of the phosphorothionate insecticides parathion and EPN by rat brain. Biochem Pharmacol 38:1597–1603, 1989.

76. P Jacob III, M Ulgen, JW Gorrod. Metabolism of (−)-(S)-nicotine by guinea pig and rat brain: Identification of cotinine. Eur J Drug Metab Pharmacokinet 22:391–394, 1997.

77. LA Peterson, A Trevor, N Castagnoli Jr. Stereochemical studies on the cytochrome P-450 catalyzed oxidation of (S)-nicotine to the (S)-nicotine delta 1′(5′)-imminium species. J Med Chem 30:249–254, 1987.

78. EF Domino. Neurobiology of PCP (Sernyl), a drug with an unusual spectrum of pharmacological activity. Int Rev Neurobiol 6:303–347, 1964.

79. EJ Holsztynska, EF Domino. Quantitation of phencyclidine, its metabolites, and derivatives by gas chromatography with nitrogen–phosphorus detection: Application for in vivo and in vitro biotransformation studies. J Anal Toxicol 10:107–115, 1986.

80. EJ Holsztynska, WW Weber, EF Domino. Genetic polymorphism of cytochrome P-450-dependent phencyclidine hydroxylation in mice. Comparison of phencyclidine hydroxylation in humans. Drug Metab Dispos 19:48–53, 1991.

81. R Herber, R Perrin, JM Ziegler, J Villoutreix, A Minn, G Siest. Identification of novel phencyclidine metabolites formed by in vitro microsomal metabolism. Xenobiotica 21:1493–1499, 1991.

82. EM Laurenzana, SM Owens. Brain microsomal metabolism of phencyclidine in male and female rats. Brain Res 756:256–265, 1997.

83. JR Crowley, PF Hollenberg. Mechanism-based inactivation of rat liver cytochrome P4502B1 by phencyclidine and its oxidative product, the iminium ion. Drug Metab Dispos 23:786–793, 1995.

84. EB Brittebo. Metabolism-dependent binding of the heterocyclic amine Trp-P-1 in endothelial cells of choroid plexus and in large cerebral veins of cytochrome P450-induced mice. Brain Res 659:91–98, 1994.

85. RJ Riley, NR Kittergham, BK Park. Structural requirements for bioactivation of anticonvulsants to cytotoxic metabolites in vitro. Br J Clin Pharmacol 28:482–487, 1989.

86. N Presle, F Lapicque, S Fournel-Gigleux, J Magdalou, P Netter. Stereoselective irreversible binding of ketoprofen glucuronides to albumin. Characterization of the site and the mechanism. Drug Metab Dispos 24:1050–1057, 1996.

87. EC Hirsch. Does oxidative stress participate in nerve cell death in Parkinson's disease? Eur Neurol 33 suppl 1:52–59, 1993.

88. AL Betz. Oxygen free radicals and the brain microvasculature. In WM Pardridge, ed. The Blood–Brain Barrier. New York: Raven Press, 1993, pp 303–321.

89. CP LeBel, SC Bondy. Oxygen radicals. Common mediators of neurotoxicity. Neurotoxicol Teratol 13:341–346, 1991.

90. B Halliwell, JM Gutteridge. Oxygen toxicity, oxygen radicals, transition metals and disease. Biochem J 219:1–14, 1984.

91. JS Beckman, TW Beckman, J Chen, PA Marshall, BA Freeman. Apparent hydroxyl radical production by peroxynitrite: Implications for endothelial injury from nitric oxide and superoxide. Proc Natl Acad Sci USA 87:1620–1624, 1990.

92. MP Murphy, MA Packer, JL Scarlett, SW Martin. Peroxynitrite: A biologically significant oxidant. Gen Pharmacol 31:179–186, 1998.

93. JX Wilson. Antioxidant defense of the brain: A role for astrocytes. Can J Physiol Pharmacol 75:1149–1163, 1997.

94. ZI Alam, SE Daniel, AJ Lees, DC Marsden, P Jenner, B Halliwell. A generalised increase in protein carbonyls in the brain in Parkinson's but not incidental Lewy body disease. J Neurochem 69:1326–1329, 1997.

95. P Riederer, E Sofic, WD Rausch, B Schmidt, GP Reynolds, K Jellinger, MB Youdim. Transition metals, ferritin, glutathione, and ascorbic acid in parkinsonian brains. J Neurochem 52:515–520, 1989.

96. W Maruyama, T Takahashi, M Naoi. (−)-Deprenyl protects human dopaminergic neuroblastoma SH-SY5Y cells from apoptosis induced by peroxynitrite and nitric oxide. J Neurochem 70:2510–2515, 1998.

97. A Bast. Is the formation of reactive oxygen species by cytochrome P450 perilous and predictable? Trends Pharmacol Sci 7:266–270, 1986.

98. TJ Monks, RP Hanzlik, GM Cohen, D Ross, DG Graham. Quinone chemistry and toxicity. Toxicol Appl Pharmacol 112:2–16, 1992.

99. MH Livertoux, P Lagrange, A Minn. Superoxide production mediated by the redox cycling of xenobiotics in rat brain microsomes: Dependence on the reduction potentials. Brain Res 725:207–216, 1996.

100. C Bayol-Denizot, JL Daval, A Minn. Xenobiotic-mediated superoxide production by cerebral cells in primary cultures (submitted).

101. S Imaizumi, T Kondo, MA Deli, G Gobbel, F Joó, CJ Epstein, T Yoshimoto, PH Chan. The influence of oxygen free radicals on the permeability of the monolayer of cultured brain endothelial cells. Neurochem Int 29:205–211, 1996.

102. GT Gobbel, TYY Chan, PK Chan. Nitric oxide- and superoxide-mediated toxicity in cerebral endothelial cells. J Pharmacol Exp Ther 282:1600–1607, 1997.

103. P Lagrange, IA Romero, A Minn, PA Revest. Transendothelial permeability changes induced by free radicals in an in vitro model of BBB. Free Radical Biol Med 27:667–672, 1999.

104. DG Graham. Oxidative pathways for catecholamines in the genesis of neuromelanin and cytotoxic quinones. Mol Pharmacol 14:633–643, 1978.

105. AN Basma, EJ Morris, WJ Nicklas, HM Geller. L-Dopa cytotoxicity to PC12 cells in culture is via its autoxidation. J Neurochem 64:825–832, 1995.

106. CL Li, P Werner, G Cohen. Lipid peroxidation in brain: interactions of L-DOPA/dopamine with ascorbate and iron. Neurodegeneration 4:147–153, 1995.

107. A Kastner, EC Hirsch, O Lejeune, F Javoy-Agid, O Rascol, YAgid. Is the vulnerability of neurons in the substantia nigra of patients with Parkinson's disease related to their neuromelanin content? J Neurochem 59:1080–1089, 1992.

108. TJ Montine, DB Farris, DG Graham. Covalent crosslinking of neurofilament proteins by oxidized catechols as a potential mechanism of Lewy body formation. J Neuropathol Exp Neurol 54:311–319, 1995.

109. TJ Montine, MJ Picklo, V Amarnath, WO Whetsell Jr, DG Graham. Neurotoxicity of endogenous cysteinylcatechols. Exp Neurol 148:26–33, 1997.

110. WR Gibb, AJ Lees. The significance of the Lewy body in the diagnosis of idiopathic Parkinson's disease. Neuropathol Appl Neurobiol 15:27–44, 1989.

111. JP Spencer, A Jenner, J Butler, OI Aruoma, DT Dexter, P Jenner, B Halliwell. Evaluation of the pro-oxidant and antioxidant actions of L-DOPA and dopamine in vitro: Implications for Parkinson's disease. Free Radical Res 24:95–105, 1996.

112. RS El-Bachá, A Minn, P Netter. Toxic effects of apomorphine on rat glial C6 cells: Protection with antioxidants. Biochem Pharmacol (in press).

113. SK Han, C Mytilineou, G Cohen. L-DOPA up-regulates glutathione and protects mesencephalic cultures against oxidative stress. J Neurochem 66:501–510, 1996.

114. M Gassen, A Gross, MB Youdim. Apomorphine enantiomers protect cultured pheochromocytoma (PC12) cells from oxidative stress induced by H_2O_2 and 6-hydroxydopamine. Mov Disord 13:242–248, 1998.

115. JJ Hagan, DN Middlemiss, PC Sharpe, GH Poste. Parkinson's disease: Prospects for improved drug therapy. Trends Pharmacol Sci 18:156–163, 1997.

116. J Lyytinen, S Kaakkola, S Ahtila, P Tuomainen, H Teravainen. Simultaneous MAO-B and COMT inhibition in L-Dopa-treated patients with Parkinson's disease. Mov Disord 12:497–505, 1997.

117. P Martinez-Martin, CF O'Brien. Extending levodopa action: COMT inhibition. Neurology 50:S27–S32, 1998.

118. C Mytilineou, SK Han, G Cohen. Toxic and protective effects of L-Dopa on mesencephalic cell cultures. J Neurochem 61:1470–1478, 1993.

119. M Saran, C Michel, W Bors. Radical functions in vivo: A critical review of current concepts and hypotheses. Z Naturforsch [C], 53:210–227, 1998.

120. CK Sen. Redox signaling and the emerging therapeutic potential of thiol antioxidants. Biochem Pharmacol 55:1747–1758, 1998.

121. ME Ginn-Pease, RL Whisler. Redox signals and NF-kappaB activation in T cells. Free Radical Biol Med 25:346–361, 1998.

10

Targeting Macromolecules to the Central Nervous System

Ulrich Bickel
Texas Tech University, Amarillo, Texas

I. INTRODUCTION

The era of high throughput screening in drug research has not yet changed the basic approach to the development of neuropharmaceuticals as far as the aspect of delivery to the central nervous system as the target organ is concerned. Like "classical" neuropharmaceuticals, which were found by trial and error, most neuroactive drugs in clinical use are typically small molecular weight compounds (M W < 600–800) with a sufficient degree of lipophilicity. These structural characteristics allow for diffusion-mediated, passive penetration through the blood–brain barrier (BBB), the morphological substrate of which is the luminal and abluminal plasma membrane of brain capillary endothelial cells, that is, a double lipid bilayer separated by the endothelial cytosol. There are few noteworthy exceptions to this general rule, where drug delivery to the brain relies on specific carrier mechanisms present at the BBB instead of diffusion. The most prominent example is the therapeutic use of L-Dopa in Parkinson's disease. Unlike the neurotransmitter dopamine, which cannot cross the BBB in significant amounts, its precursor L-Dopa is a substrate for the transporter of large, neutral amino acids (1). Thus, L-Dopa therapy is an example of rational drug design based on knowledge of BBB transport biology. There are a number of other small molecular weight drugs on the market and under development that make use of physiological carrier mechanisms for entry into the CNS compartment (see Chapter 8 of this volume).

Despite these successful examples of targeted drug delivery to the CNS, the employment of preexisting uptake mechanisms at the BBB for development of small molecule therapeutics still may be regarded as optional. In contrast, any attempt to exploit the pharmacologic potential of macromolecules invariably requires the incorporation of a delivery strategy. The spectrum of future neuropharmaceuticals falling into that category ranges from peptide-based drugs (neuroactive peptides, neurotrophic factors, cytokines, monoclonal antibodies) to nucleotide-based agents such as antisense oligodeoxynucleotides (ODNs) and genes, and the number of attractive drug candidates is rising. At present both the discovery of new active principles by molecular biological approaches and their subsequent characterization in reductionist in vitro systems is progressing at a rapid rate. For example, more than 30 proteins with neurotrophic activity have been described to date, which could become valuable therapeutic agents in a variety of diseases affecting the CNS from ischemic brain damage to Alzheimer's disease (2). In comparison to these advances, resources devoted to the development of delivery strategies fall short. Nevertheless, promising drug delivery strategies for macromolecules are presently under active investigation. Since diseases afflicting the CNS are diverse in etiology (infectious, degenerative, autoimmune, metabolic, tumors), severity, and time course, therapeutic interventions have a corresponding range of targets and aims. It can therefore be predicted that there will be not a single solution to the delivery of pharmacological agents. Based, on their principal characteristics, drug delivery strategies can be broadly classified into the categories local versus systemic, invasive versus noninvasive, and biochemical/pharmacological versus physiological. This chapter emphasizes the physiological approach but begins with an overview of the other options.

II. INVASIVE DELIVERY STRATEGIES

A. Pharmacokinetic Aspects of Intracerebroventricular and Intracerebral Drug Administration

Invasive approaches to drug local delivery require (neuro)surgical intervention. This includes methods that physically bypass the BBB by direct delivery into the cerebrospinal fluid or brain parenchyma. Intrathecal and intracerebral drug administration differs fundamentally from systemic drug administration in terms of pharmacokinetic characteristics determining brain tissue concentrations: in that case the available dose reaching the target organ is 100%. However, there are large gradients inside the tissue with very high local concentrations at the site of administration (the ventricular surface or tissue site of injection) and zero concentration at some distance. Because macromolecules have low diffusion coefficients, the gradients will be even steeper than what has been measured for small molecular weight drugs (3). After intracerebroventricular (icv) injection, the rate of elimination from the CNS compartment is dominated by cerebrospinal fluid dynamics. In humans, the entire cerebrospinal volume is exchanged every 4–5 hours (4). Drugs in solution may be administered by single injection or long-term infusion through implanted catheters. Since CSF exchange by bulk flow limits the half-life of any substance injected into CSF, a long-term infusion strategy is mostly adopted in studies involving icv application.

Clinical examples of intrathecal small drug delivery are the icv administration of glycopeptide and aminoglycoside antibiotics in meningitis (5), the intraventricular treatment of meningeal metastasis (6), the intrathecal injection of baclofen for treatment of spasticity (7), and the infusion of opioids for severe chronic pain (8). These examples have in common that the drug targets in all instances are close to the ventricular surface. Superficial targets may also be accessible for some macromolecular drugs. In the case of nerve growth factor (NGF), specific receptors are expressed on axons running in the fimbria–fornix (9). Tracer pharmacokinetic studies showed that the direct tissue penetration of NGF after intraventricular injection is marginal and does not extend deeper than 1–2 mm from the surface of the infused ventricle in rats (10) and beyond 2–3 mm in the primate brain (11). However, retrograde transport of labeled NGF to neuronal cell bodies in cholinergic basal forebrain nuclei occurred after icv injection. Correspondingly, beneficial effects of NGF on the survival of cholinergic neurons could be demonstrated in lesion models in rodents (12) and primates (13). These studies supported the speculation that icv NGF could be a treatment in Alzheimer's disease (AD) and other neurodegenerative disease states involving the cholinergic system. To date the published reports of NGF administered by the ventricular route to treat AD have involved only a small number of patients (14). In this study, in the absence of significant beneficial effects, side effects such as significant weight loss and constant back pain were noticed. It was concluded that intracerebroventricular administration is unsuitable for the delivery of NGF (15). In addition, earlier animal studies had indicated the induction of hyperinnervation of cerebral blood vessels following icv infusions as a potential untoward effect (16).

With other growth factors, such as brain-derived neurotrophic factor (BDNF), retrograde transport mediated by receptors in the vicinity of the ventricular surface appears much more restricted, which makes intraventricular administration even less promising (17).

Restricted diffusion also limits tissue distribution after intraparenchymal drug administration. Distribution has been measured in the rat brain after implantation of polymer discs containing NGF (18, 19). Drug concentrations decreased to less than 10% of the values measured on the disc surface within a distance of 2–3 mm, even after prolonged periods (several days). Therefore, applying this approach in the large human brain would require the repetitive stereotaxic placement of multiple intraparenchymal depots. The same pharmacokinetic limitation is true in principle for the implantation of encapsulated genetically engineered cells (20), which synthesize and release neurotrophic factors.

A special case is drug distribution caused by convective flow under high flow microinfusion, and its potential to overcome limited diffusion by convective flow was tested in animal experiments by Morrison et al. (21). Intraparenchymal high flow microinfusions with flow rates up to 4 μl/min were shown to result in almost homogeneous tissue concentrations of macromolecules (transferrin, MW 80 kDa) over a large volume and over a distance of exceeding 10 mm from the catheter tip within an infusion period of 2 hours (21, 22). The flat concentration profile in the zone of convective flow is in contrast to the steep concentration gradients associated with diffusion-mediated distribution. The method has been transferred to clinical trials for treatment of gliomas and metastatic brain tumors with the targeted toxin Tf-CRM107 (23), a chemical conjugate of human transferrin with the mutant diphtheria toxin CRM107; the transferrin targets the toxin to proliferating cells that highly express the transferrin receptor. CRM107 has selectively lost the intrinsic binding affinity to mammalian cells by a point mutation.

B. Drug Delivery Based on Blood–Brain Barrier Disruption

Another invasive strategy for drug delivery to the brain is the temporary physicochemical disruption of endothelial integrity. Experimental barrier opening for low molecular weight tracers and macromolecules (Evans blue–albumin) was demonstrated with intracarotid infusions of membrane active agents like bile salts (24), oleic acid (25), the cytostatic drugs etoposide (26) and melphalan (27), and cytochalasin B (28). Intracarotid low pH buffer infusion also opens the BBB (29).

1. Hyperosmolar Barrier Opening

Most studies, however, are available with hyperosmolar solutions, a principle that was described by Broman (30) and extended by Rapoport (31). Technically, the procedure requires general anesthesia and high flow, short-term infusion of 25% mannitol or arabinose. Hypertonic disruption is under clinical evaluation for enhanced delivery of small molecular weight cytostatic agents to brain tumors (32). The underlying mechanism is a sequela of endothelial cell shrinkage, disruption of tight junctions, and vasodilation by osmotic shift. Morphological studies in rats, where the induction of neuropathological changes by osmotic opening was examined, provided evidence of brain uptake of macromolecules: the extravasation of plasma proteins such as fibrinogen and albumin was shown immunohistochemically at the light microscopic level (33). Electron microscopy also revealed ultrastructural changes such as swelling of astrocytic processes and severe mitochondrial damage in neurons (34). Infusion of albumin–gold complexes after BBB disruption by intracarotid hyperosmolar arabinose was used to visualize the cellular mechanism in rats at the electron microscopic level (35). In addition to opening of junctional complexes and the formation of interendothelial gaps, transendothelial openings and tracer passage through the cytoplasm of injured endothelial cells were observed. In response to hyperosmotic barrier disruption, there was also evidence of prolonged (24 h) cellular stress or injury in neurons and glia, as expressed by the induction of heat shock protein (HSP-70) (36).

It has been shown that the barrier opening for high molecular weight compounds is of shorter duration than that for small molecules (37). When the degree of barrier opening is measured with methods that are suitable for a regional evaluation (autoradiography in animal studies, positron emission tomography in man), there is a characteristic difference in the degree of barrier opening in the tumor versus normal brain. This opening was consistently found to be more pronounced for the normal BBB (38, 39). While the nonspecific opening of the BBB to plasma proteins has a potential to elicit neuropathological changes, osmotic disruption has been tested as a strategy for the brain delivery of macromolecular drugs such as monoclonal antibodies, nanoparticles, and viruses. Quantitative uptake studies after hyperosmolar BBB opening in animals and humans were performed with radiolabeled monoclonal antibodies and their antigen binding fragments against various tumor antigens (40–42). In normal rat brain, a 25- to 100-fold relative accumulation in the BBB-disrupted hemisphere of a radioiodinated rat monoclonal antibody (IgM) against human small-cell carcinoma of the lung was reported (40). The mean PS value at 10 minutes after BBB opening and intracarotid antibody infusion was calculated as 8.36×10^{-6} s^{-1} (= 0.5 µl min^{-1}g^{-1}). In patients with intracranial melanoma metastasis, ^{131}I-labeled antigen binding fragments of melanoma-specific antibodies were infused intravenously in conjunction with BBB disruption (41). Brain uptake was measured by gamma camera imaging and was used to calculate PS values at 3 hours. A mean PS value of $1.16 \times$

10^{-6} s^{-1} was estimated for the treated hemisphere, which was threefold higher than for the nonperfused hemisphere. Owing to the transient nature of BBB opening, the calculated PS values might represent only rough estimates. At any rate, no specific enhancement of tracer uptake in tumor versus normal brain was seen in the patient study (41). The ability of osmotic disruption to deliver 20 nm iron oxide particles to normal brain was postulated in another study (43). Similarly, recombinant adenovirus or herpesvirus was delivered by intracarotid administration to normal brain tissue (44) and to tumor xenografts in nude rats (45).

2. Biochemical Barrier Opening

BBB opening may also be achieved by receptor-mediated mechanisms. The vasoactive compounds prostaglandins, histamine, serotonin, leukotriene C4 (LTC4), and bradykinin have all been shown to induce BBB leakage (46). The effects of LTC4 and bradykinin are more pronounced on the blood–tumor barrier than on the normal BBB (47, 48). In the case of LTC4, that difference is ascribed to the presence of an enzymatic barrier in normal brain tissue due to the endothelial expression of γ-glutamyltransferase (γ-GT). The enzyme metabolizes and inactivates LTC4 to LTD4 (49). In contrast, tumor vessels are unable to express equivalent activities of γ-GT, a fact that may be exploited for selective opening of the tumor barrier by intracarotid administration of LTC4. However, the effect is restricted to small molecules, as indicated by the absence of any increase in the tumor accumulation of a dextran tracer of molecular weight 70 kDa (47). On the other hand, bradykinin opens the barrier in the high molecular weight range, too. It acts on endothelial cells through B$_2$ receptors located on the abluminal side. Normal brain tissue is protected from barrier opening by bradykinin in the vascular lumen because the peptide cannot access these receptors. In tumor vessels the barrier integrity is sufficiently compromised to allow for a bradykinin-mediated additional opening at low peptide concentrations (48). The effect shows a rapid desensitization within 60 minutes. An increase in intracellular Ca^{2+} concentration has been shown, which in turn transiently disrupts intercellular tight junctions (50). In addition, the nitric oxide–cyclic GMP pathway is involved. While bradykinin itself requires intracarotid administration, an analogue with prolonged half-life (RMP-7) is effective after intracarotid (51) or intravenous (52) application. The drug is evaluated in the therapy of human malignant gliomas to enhance delivery of carboplatin to the tumor. Recently, a four- to fivefold increase in the delivery of the cytokines interferon γ, tumor necrosis factor α, and interleukin 2 to experimental brain tumors (RG2 glioma) after intracarotid infusion of RMP-7 was demonstrated (53).

III. PHYSIOLOGICAL DELIVERY STRATEGIES

A. Pharmacokinetic Aspects of Systemic Drug Administration

The desirable pharmacokinetic characteristics of a given macromolecular drug can be defined only within the context of the individual application. This may be exemplified by two situations: for radioimmunodetection of a brain tumor using a specific antibody to a tumor antigen, the signal-to-noise ratio, or tissue-to-plasma ratio, of the radiolabeled antibody is more important than the absolute activities achieved in tumor tissue. In contrast, if the same antibody were used to deliver a radioisotope, a toxin, or a cytostatic agent as a therapeutic substance, the absolute amount accumulating in the tissue over time would be the primary parameter of interest (54).

Following systemic drug administration, uptake from the circulation into parenchyma by a specific organ of interest will be determined by the following factors: (a) blood flow to the organ, (b) permeability of the microvascular wall, and (c) the amount of drug available for uptake, which is inversely related to systemic clearance and is represented by the area under the plasma concentration–time curve (AUC). Focusing on the brain, pharmacokinetics of macromolecules represents the case in which the extraction rate from blood plasma into the organ is not limited by blood flow (a valid assumption as long as the extraction during single capillary passage is below 20%, Ref. 55). Then, blood flow may be neglected as a parameter in calculating tissue uptake. For the quantification of brain tissue accumulation (C_{brain}) at time T during the phase of unidirectional uptake, the following expression holds:

$$C_{brain}(T) = \text{PS} \times \text{AUC}|_0^T$$

where PS is the brain capillary permeability surface area product, an expression equivalent to the organ clearance, and AUC is the area under the plasma concentration time curve. It should be mentioned that this equation does not take into account efflux of either intact drug or metabolization and efflux of degradation products from brain. Measurement of efflux is covered in detail in (Chapter 6 of this volume).

B. Physiological Transport Mechanisms for Peptides and Proteins at the Blood–Brain Barrier

As opposed to the delivery strategies discussed above, which are primarily aimed at short-term application in the treatment regimens of malignant brain tumors, drug treatments of chronic degenerative disorders will require long-term application of the therapeutic agent. This implies the need to develop a noninvasive approach for brain delivery via the systemic route. To this end, utilization of endothelial transport mechanisms is being explored in preclinical studies. For macromolecules, the uptake mechanism required for transendothelial passage is necessarily mediated by vesicular transport, not by a passage through pores. At this point, the analogy to the "pseudonutrient" approach, which utilizes nutrient transporters, such as the example with L-Dopa, is limited. In carrier-mediated uptake, the drug or prodrug that is targeted to receptors within the central nervous system is also a substrate for specific transport proteins at the BBB. The structural variations of compounds suitable for delivery are limited to the degree that is tolerated by the corresponding BBB carrier protein. Therefore, only small molecular weight drugs can exploit carrier-mediated transport. In contrast, receptor-mediated endocytosis of a peptide or protein ligand does not have the narrow size restrictions of carrier-mediated uptake through pores in the plasma membrane.

1. Receptor-Mediated Uptake

The concept of saturable, receptor-mediated uptake systems for peptides and proteins at the blood–brain barrier has evolved over the last two decades. There is now combined evidence from in vitro and in vivo studies, at the biochemical and pharmacokinetic levels as well as at the morphological level, that peptides are transported across the endothelial cells. This includes transport of compounds as structurally diverse as insulin and insulin-like growth factors (IGF-I and II) (56), transferrin (57), low density lipoprotein (LDL) (58), and leptin (59). The overall process of transendothelial passage is designated as transcytosis and is composed of binding to a luminal plasma membrane receptor, endocytosis, transfer through the endothelial cytoplasm to the abluminal side, and abluminal exocytosis into brain interstitial space (60).

The binding of insulin at the BBB is mediated by the insulin receptor α-subunit as demonstrated by affinity cross-linking of [^{125}I]insulin to isolated human brain capillaries. Gel electrophoresis (SDS-PAGE) of the solubilized receptor revealed a band corresponding to the 130–135 kDa molecular weight expected for the glycosylated α-subunit (61). This finding fits the results of radioligand binding assays with isolated cerebral microvessels from different species including man (61–64), which showed specific binding and internalization of insulin. Endocytosis could be verified by the demonstration that a nonsaturable fraction of approximately 75% of the capillary binding at 37°C was resistant to a mild acid wash (61). Very similar data were obtained in primary cultures of bovine brain microvascular endothelial cells (65). These in vitro results corresponded to the measurable in vivo brain uptake of insulin under intracarotid infusion (66). In these experiments degradation of the tracer was excluded by high performance liquid chromatographic (HPLC) analysis, and evidence of parenchymal uptake beyond the vascular wall was shown by thaw-mount autoradiography of cryosectioned brain slices.

A similar set of in vitro and in vivo data attests to the expression of IGF receptors at the BBB and transport of their ligands (67–69). Apparently, there are species-specific differences in this system, since the human BBB expresses predominantly the type III IGF receptor (68), while the type II IGF receptor, which is identical to the mannose 6-phosphate receptor, is absent. In the rat, presence of IGF type I and type II receptors at the BBB has been described on the basis of receptor binding assays and in situ hybridization (67, 69).

Following the demonstration of high levels of transferrin receptor expression on rat brain microvessels with a specific monoclonal antibody (70), transferrin binding to isolated human brain microvessels was shown in radioreceptor studies. A saturable, time-dependent binding with a dissociation constant K_D of 5.6 nM was found (71). Subsequently, the transport of transferrin through the BBB was measured in vivo. While it is obvious that transferrin is involved in the delivery of iron to the endothelial cell, there is no agreement yet in the literature on the extent to which the exocytosis of iron into brain interstitial fluid occurs in a transferrin-bound mode. Fishman et al. (57) and Skarlatos et al. (72) have reported experiments in support of significant transcytosis of the 80 kDa plasma iron transport protein. These studies employed ^{125}I-labeled transferrin tracer and brain perfusion in the rat; that is, the methods applied allow for the control of transferrin concentration by avoiding admixture of endogenous plasma. This is crucial because of the high concentration of transferrin in plasma of about 25 µM. Therefore, the BBB transferrin receptor is saturated under physiological conditions. A 90% inhibition of the uptake was found in the presence of 10% normal rat serum in the perfusate (72), which explains the results after intravenous administration of [^{59}Fe-^{125}I]-transferrin, where only a spurious tracer uptake in brain was found (73). In 1996, however, transcytosis of transferrin from the apical (luminal) to the basolateral (abluminal) surface also was demonstrated in an in vitro BBB model, coculture of bovine brain endothelial cells and rat astrocytes (74). There was saturable and temperature-sensitive transport of [^{125}I]holotransferrin, with measurable transport at 37°C but not at 4°C, and no transport of iron-depleted [^{125}I]apotransferrin, which has low affinity to transferrin receptors. Moreover, when double-labeled [^{59}Fe-^{125}I]transferrin was used as a tracer, the transport of iron was found to be twice as high on a molar basis as the transport of transferrin.

That ratio corresponds to the two binding sites of transferrin for iron and provides evidence for cotransport. Another concern expressed with regard to transcytosis of transferrin at the BBB is the failure to detect TfR on the abluminal plasma membrane of endothelial cells (75) when a "preembedding" approach at the electron microscopy level was employed. However, the detection of abluminal antigens in electron microscopy requires the application of post-embedding techniques, which in turn involves initial tissue fixation, associated with the potential loss of immunoreactivity.

This methodological problem was recently addressed in a confocal microscopy study with freshly isolated rat brain capillaries (76). The method permits omitting of the fixation step. Fluorescent immunoliposomes were synthesised by attachment of the anti-TfR antibody OX26 to pegylated liposomes carrying rhodamine-phosphatidylethanolamine. The high fluorescence intensity of the liposomes permitted the full exploitation of the spatial resolution of confocal microscopy, and it was possible to demonstrate the presence of TfR unequivocally on both the luminal and abluminal plasma membranes of endothelial cells. In addition, the pattern of immunofluorescence was compatible with an intracellular accumulation of OX26 liposomes in endosomal structures. These data fit well with the staining pattern seen by confocal microscopy in endothelial cell monolayers after incubation with fluorescein-conjugated holotransferrin (74). Luminal binding of the OX26 TfR antibody, its endocytosis, accumulation in endosomes and multivesicular structures, and abluminal exocytosis have been observed at the electron microscopic level after in vivo infusion of the monoclonal antibody conjugated with 5 nm colloidal gold (77). Figure 1 depicts these crucial steps.

Recently, substantial evidence has been accumulated to support the presence at the BBB of a transport system that is involved in the transcytosis of leptin. A short cytoplasmatically truncated leptin receptor isoform was first cloned from the choroid plexus (i.e., the site of the blood–CSF barrier). Subsequently it was shown that leptin in plasma enters brain tissue in mice by a saturable mechanism (59). The specific binding of leptin to a high affinity site ($K_D = 5.1 \pm 2.8$ nM) could be demonstrated with isolated human brain capillaries, which also internalized the ligand at 37°C (78). At the mRNA level, in situ hybridization and evidence acquired by means of the reverse transcriptase polymerase chain reaction showed that brain microvessels express even higher amounts of the short receptor isoform message than choroid plexus (79). These receptors are also subject to regulation, inasmuch as rats on a chronic high fat diet express higher amounts of the corresponding mRNA and protein in their brain capillaries (80).

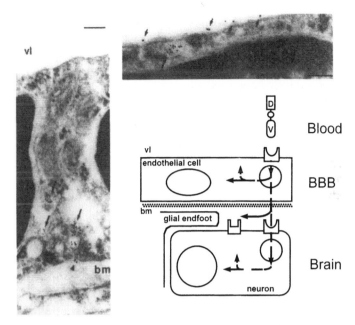

Fig. 1 Steps in the transcytosis of the TfR antibody OX26 through a brain capillary endothelial cell are shown in the electron micrographs. (From Ref. 77.) The antibody was conjugated to 5 nm colloidal gold particles and was infused into the internal carotid artery of rats. Binding of the antibody to the luminal plasma membrane is indicated by short arrows (top). The long arrows mark clusters of internalized antibodies inside vesicular structures (left). Abluminal exocytosis is indicated by the arrowhead. Scale bar = 100 nm. The scheme on the right depicts transendothelial chimeric peptide delivery. The receptor on the luminal plasma membrane binds the vector moiety and mediates endocytosis. Pharmacological effects have been shown for peptide drugs acting on a cognate plasma membrane receptor on brain cells that is specific for the drug moiety (see Sec. III.C.3). An intracellular drug effect (e.g., by antisense mechanisms) requires release of the drug from endosomal vesicles. Release may occur inside endothelial cells or inside brain cells. The latter demands another receptor mediated internalization. Abbreviations: vl, vascular lumen; bm, basement membrane; V, vector; D, drug.

Evidence in support of the notion that the transcytotic pathway can accommodate rather large "payloads" comes from in vitro studies with lipoproteins and endothelial monolayers. The size of LDL particles ranges from about 15 to 25 nm. In contrast to chemically modified lipoproteins like acetylated LDL, which is taken up by brain capillary endothelial cells in vitro and in vivo by endocytosis but is not translocated into brain (81), native LDL undergoes apical-to-basolateral transport from the blood to the brain side of bovine brain endothelial cells in primary cultures (58). Also, a modulation of the endothelial transport in this system by cocultured astrocytes was described.

2. Absorptive-Mediated Uptake of Lectins and Cationic Peptides/ Proteins

A mechanism of brain uptake that is related to receptor-mediated transcytosis operates for peptides and proteins with a basic isoelectric point ("cationic" proteins) and for some lectins (glycoprotein-binding proteins). The initial binding to the luminal plasma membrane is mediated by electrostatic interactions with anionic sites, or by specific interactions with sugar residues, respectively, and the transport is termed "adsorptive-mediated transcytosis." Ultrastructural studies utilizing enzymatic treatment and lectins coupled to colloidal gold revealed that anionic sites and carbohydrate residues exhibit a polarized distribution on the luminal and abluminal membranes (82): negative charges are more abundant on capillaries than on arterioles or venules, and the luminal surface expresses glycoproteins with sialic acid residues, while the abluminal membrane carries heparan sulfates.

Morphologic evidence of transcytosis after intracarotid infusion of cationized polyclonal bovine immunoglobulin was seen by autoradiography at the light microscopic level (83). At the electron microscopic level, conjugates of horseradish peroxidase and wheat germ agglutinin (WGA-HRP) labeled the abluminal subendothelial space of brain microvessels after intravenous administration (84). Uptake of various natural and chemically modified basic proteins through the BBB has been measured in numerous pharmacokinetic studies—for example, with histones (85), recombinant CD4 (86), avidin (87), cationized albumins (88, 89), cationized polyclonal IgG (83, 90, 91), and various cationized monoclonal antibodies (92, 93). As shown in studies by Terasaki et al. (94) for the heptapeptide E-2078, small basic peptides are able to undergo adsorptive-mediated transport, too.

Native or recombinant proteins (e.g., albumin, antibodies, growth factors) can be chemically derivatized by the introduction of amine groups on accessible carboxyl side chains (95). Activation of these carboxyl groups by carbodiimide reagents is followed by coupling of hexamethylenediamine or naturally occurring polyamines like putrescine, spermine, and spermidine (96). Care must be taken not to compromise the biological activity of the cationized proteins. The applicability of site protection was demonstrated in the case of a monoclonal antibody, AMY33, which is directed against a synthetic peptide representing amino acids 1–28 of the β-amyloid peptide of Alzheimer's disease. Cationization of the antibody in the presence of a molar excess of the specific peptide antigen prevented significant loss of binding affinity (97).

In quantitative terms, the adsorptive mechanism is distinguished from receptor-mediated uptake by lower affinity and higher capacity (98, 99). In theory, this could result in comparable overall transport rates through the BBB. In practice, the measured brain concentrations (percent injected dose per gram: %ID/g) for cationic proteins may be limited by the fact that cationized proteins show profoundly increased uptake into organs other than brain, predominantly liver and kidney (99). Widespread tissue uptake is equivalent to an enhanced systemic clearance and lower AUC, thereby limiting the amount of drug available for BBB transport. When the organ distribution of different cationized proteins is compared, varying degrees of accumulation are found in some organs (e.g., liver uptake of cationized immunoglobulins is much higher than that of cationized albumin) (99, 100).

Structure–activity relationships for the brain uptake of a small tetrapeptide have been presented (101). The authors found that basicity and C-terminal structure were important determinants for endothelial endocytosis of the synthetic peptide 001-C8 (H-MeTyr-Arg-MeArg-D-Leu-NH(CH$_2$)$_8$NH$_2$). The situation is certainly more complex for large proteins. When the brain uptake of superoxide dismutase (SOD) bearing modifications with either putrescine, spermine, or spermidine was measured, the highest PS product was found for the putrescine derivative, which has the lowest number of cationic charges (96) pointing to additional factors beyond electrostatic interactions. It remains to be seen whether detailed analyses of structural requirements for initiation of adsorptive-mediated transcytosis will identify cationic modifications that allow targeting to the vascular bed of an organ.

Toxicological and immunological consequences of long-term administration of cationized albumin have been addressed (88). It could be shown that under repetitive administration of the homologous protein (i.e., cationized rat albumin used in rats) there was no organ toxicity or deviation in blood chemistry detectable compared to a control group receiving native rat albumin. Apparently, homologous proteins are tolerated after cationization without causing the immunologic reactions and organ damage (e.g., deposition of immune complexes in the glomerulus of rats after treatment with cationized bovine albumin) found with heterologous cationized proteins (102).

The general property of cationic proteins to escape vascular barriers could also be utilized for the delivery of radiopharmaceuticals to tumors or metastasis throughout the body (93, 103) and for the treatment of viral infection with cationized antibodies (91).

C. Drug Delivery by Chimeric Peptides

The opportunity for drug delivery arises from the possibility to synthesize "chimeric peptides" (60). These are generated by linking a drug that lacks transport at the BBB to a vector. Binding of the vector at the luminal membrane of brain capillary endothelial cells initiates receptor-mediated or adsorptive-mediated transcytosis. The mode of delivery was schematically visualized in Fig. 1. Size and structure of the cargo may vary as long as the drug moiety does not inhibit binding and cellular uptake of the vector and may be limited only by the size of endocytotic compartments. Figure 2 depicts the chimeric peptide concept in a three-dimensional arrangement to give an impression of its multiple variations. These may be tailored for an individual application.

In the initial demonstrations of brain delivery based on the chimeric peptide strategy, the vector was cationized albumin and the peptide drug directly coupled to it was the opioid peptide β-endorphin (95) or its metabolically stabilized analogue [D-Ala^2]β-endorphin (104). These tracer studies used a chimeric peptide labeled in the endorphin moiety and provided evidence of internalization by isolated brain capillaries and transport into brain tissue in vivo.

Quantitative measurements of the uptake of cationized proteins as vectors in chimeric peptides were subsequently performed with cationized human albumin conjugated to avidin. Vectors utilizing receptor-mediated uptake must avoid competition by endogenous ligands, as discussed for transferrin above, and also should not display undesirable intrinsic pharmacological activity. Therefore, the use of insulin, with its effect as a hypoglycemic hormone, would be undesirable. Despite these caveats, experimental evidence in favor of the potential utility of insulin peptides in a delivery system has been presented with an insulin fragment: one tryptic fragment with low receptor binding affinity (10% of the affinity of insulin) was devoid of a hypoglycemic effect in mice in vivo, yet a chimeric peptide synthesized with horseradish peroxidase was transported into brain in vivo and reached a brain concentration of 1.41% of the injected dose after intravenous administration to mice (105).

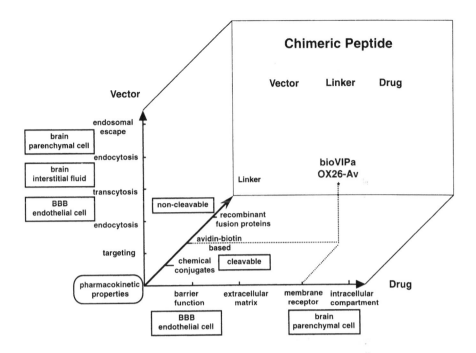

Fig. 2 The qualities of a chimeric peptide depend on each of its domains. The vector moiety provides targeting and transport to brain endothelial cells and beyond. Drugs could be designed to act at the level of the BBB, to a target in extracellular space, or on brain cells. Intracellular effects (e.g., of antisense-ODN) require the incorporation of release mechanisms into the construct. The different options for the linker domain add more complexity. Overall, the pharmacokinetic properties of the chimeric peptide are of paramount importance. The chimeric peptide modeled is bio VIPa/OX26-Av, made up from a biotinylated VIP analogue and a conjugate of OX26 with avidin; it is an example of a chimeric peptide that elicits a pharmacological effect by binding to cell surface receptors after transcytosis through the BBB.

An alternative approach, however, utilizes vectors based on monoclonal antibodies specific to the extracellular domain of a peptide or protein receptor at the BBB; such vectors fulfill the criteria of binding to the receptors at a site distinct from the ligand binding site (noncompetitive), and not interfering with the endocytosis process.

1. Pharmacokinetic Evidence for the Brain Uptake of Chimeric Peptides

Table 1 compares pharmacokinetic data accumulated for some vectors (135). Quantitative comparisons within the same species are possible for the rat regarding vectors derived from the anti-TfR monoclonal antibody OX26 and from cationized human serum albumin. To put the efficiency of brain delivery into perspective, the comparison to a classical neuroactive drug may be informative. In the rat, morphine does not reach brain concentrations in excess of 0.08%ID/g following systemic administration (106). The vector cHSA-NLA, based on cationized albumin, accumulates to approximately the same level within 1 hour (Table 1) and reaches maximum concentrations of 0.15 %ID/g at 6 hours (87), while OX26 after one hour already reaches concentrations in rat brain that are three to four times higher than morphine (Table 1; Ref. 107). The difference in terms of brain concentrations between vectors based on OX26 compared to the vectors based on cationized human albumin is mainly due to a corresponding difference in the PS products; that is, the rate of uptake by absorptive-mediated transcytosis of cationized albumin at the BBB is lower than the rate of receptor-mediated uptake of OX26.

With regard to transport capacity, the introduction of the human insulin receptor antibody (HIRMAb) 83-14 as a vector (108) indicates the potential for future improvements in brain-specific delivery vectors. Compared to anti-TfR monoclonal antibodies, the brain delivery in primates is over seven-fold higher, and this is due to the high PS product of the HIRMAb. The saturable character of receptor-mediated processes needs to receive attention as well, and full characterization of the transport capacity for drug delivery requires investigation of dose dependence. Noncompeting antibodies such as OX26 avoid the problem of competition by endogenous ligand; however, there remains the saturability of the antibody binding site. The values in Table 1 were obtained in uptake experiments with doses of vectors in the low micrograms-per-kilogram range, corresponding to plasma concentrations in the low nano-molar range. Linear pharmacokinetics cannot be expected at higher doses, which result in changes in both plasma AUC and apparent PS product. An example was provided in a dose escalation study with coinjection of OX26-avidin and increasing amounts of OX26 up to 2 mg/kg (109), where an increase in AUC and a corresponding decrease in PS product were seen. In that case brain drug delivery, expressed as percent injected dose per gram, stayed at an almost constant level because the saturation of systemic clearance offset the decrease in transport rate at the BBB with higher doses.

Table 1 Brain Concentration, Blood–Brain Barrier PS Product, and Plasma AUC (0–60 min) of Brain Delivery Vectors After Intravenous Bolus Injection in Rats

Vector	Brain Concentration (%ID/g)	PS (μl min^{-1} g^{-1})	AUC, 0–60 min (%ID min ml^{-1})
^{3}H-OX26	0.27 ± 0.04	1.92 ± 0.06	132 ± 19
OX26-Av[a]	0.041 ± 0.004	0.85 ± 0.02	49 ± 4
OX26-NLA[a]	0.17 ± 0.04	0.70 ± 0.10	232 ± 25
OX26-SA[a]	0.20 ± 0.03	0.92 ± 0.10	216 ± 28
cHSA-Av[a]	0.015 ± 0.006	0.26 ± 0.13	64 ± 7
cHSA-NLA[a]	0.061 ± 0.012	0.20 ± 0.04	300 ± 14
^{125}I-HIR MAb (monkey)	3.8 ± 0.4 (100 g brain)[b]	5.4 ± 0.6	5.9 ± 1.2 (0–180 min)

[a] These proteins were labeled at the avidin moiety with [^{3}H]biotin.

[b] %ID in total brain tissue after 3 hours.

Abbreviations: Av, avidin; NLA, neutral avidin; SA, streptavidin; cHSA, cationized human serum albumin; HIRMAb, human insulin receptor monoclonal antibody.

Source: Ref. 135.

2. Linker Strategies

The coupling between vector and drug moiety may be performed by means of either chemical or molecular biological approaches (Fig. 2). While the options offered by chemical methods provide rapid synthesis of conjugates, which is particularly suitable for animal experiments and "proof of concept" studies, fusion proteins have the potential for bulk production of a defined molecular entity for future clinical development.

Direct chemical conjugation of vector and drug has been applied for the coupling of small molecules (110), peptides (95), and proteins (111, 112). Avidin–biotin technology as a linker strategy (113) was adopted as an attractive alternative. The availability of biotinylating reagents for a range of compounds and functional groups provides versatility, since a single vector can be used for the delivery of different drugs. Moreover, the avidin–biotin bond is extremely stable.

The impact of pharmacokinetics necessitated a modification of the original avidin-based vectors (114). The consequences can be demonstrated by comparing two series of vector constructs, as contained in Table 1. Because avidin is highly cationic (isoelectric point > 9.3), plasma AUC of conjugates is low (secondary to uptake by liver and kidney). Replacement by neutral analogues resulted in higher plasma AUC. Both for OX26 and for cationized albumin, similar improvements in brain delivery were seen when the native avidin was replaced by chemically modified avidin (neutral avidin, NLA) or by streptavidin (SA), which is a slightly acidic bacterial analogue. Apparently, the negative effect on plasma pharmacokinetics of the avidin moiety is eliminated in IgG–avidin fusion proteins, too (115). The biotin–avidin linker strategy is particularly suitable for synthetic peptide drugs. These can be designed to facilitate monobiotinylation at a site that does not interfere with bioactivity (116–119). Monobiotinylation is required because avidin is multivalent. A 1:1 molar conjugate of vector and (strept)avidin can bind up to four biotin residues, and higher degrees of biotinylation will result in the formation of high molecular weight aggregates, which are cleared rapidly from the circulation (120).

Chimeric peptides need to be stable in the circulation before brain uptake occurs, and either amide bonds, thioether, or disulfide linkers are suitable in terms of stability in the plasma compartment (98). If binding of a peptide drug to the vector reduces binding affinity to the drug receptor on brain cells, the release of free drug in brain is required (117, 118). It was found that in vivo, disulfide-reducing enzymes in tissues can cleave in two chimeric peptides linked by disulfide (—SS—). Chromatographic analysis of whole-brain tissue from in vivo studies had demonstrated that cleavage of a biotinylated dermorphin analogue, [Lys[7]]dermorphin-amide (designated bioSSK7DA), with an N-hydroxysuccinimide-dithiopropionate (NHS-SS-) biotin linker occurs from the vector OX26-SA in brain but not in plasma (118). To further show that the chimeric peptide is stable not only in the circulation but also during transport through the BBB, a study was performed with the structurally related dermorphin analogue [^{3}H]Tyr-DArg-Phe-Lys-NH$_2$ (DALDA). DALDA was also biotinylated with NHS-SS-biotin and coupled to OX26-SA. Systemic administration resulted in a brain delivery of 0.1 %ID/g of the chimeric peptide 60 minutes after intravenous injection. Extracellular space sampling by intracranial microdialysis during this period did not reveal the presence of free peptide released from the vector by disulfide cleavage (121). The extracellular and/or intracellular compartments that are eventually responsible for disulfide cleavage of chimeric peptides within brain (118) await further characterization.

Recently the successful application of biotin-derivatized poly(ethylene glycol) (PEG) linkers with molecular weights of 2000 or 5000 Da has been reported (120, 122). Because of their length and flexibility, these linkers do not interfere with biologic activity, and they represent an alternative to cleavable linkers with shorter spacer arms.

3. Pharmacological Effects of Chimeric Peptides

The cargo that is suitable for transport by chimeric peptides encompasses a wide array of substances. Table 2, an overview of drugs that have been coupled to avidin-based vectors, also gives information on the increase in brain delivery achieved by the vector. Payloads include peptide-based therapeutics like the opioid peptide analogues DALDA (117) and K7DA (118), and an analogue of vasoactive intestinal polypeptide (VIP) (116, 119). Such peptide drugs have in common that their target is a specific receptor on the plasma membrane of brain cells projecting into extracellular space (see also scheme in Fig. 1). Similarly, high molecular weight proteins like neurotrophic growth factors target plasma membrane receptors (111, 123). The examples of chimeric peptides with proven pharmacological activity in vivo all belong to that category.

Table 3 summarizes studies that measured CNS effects after peptide drug delivery. VIP is suitable for the demonstration of a pharmacological effect with a vector-mediated drug delivery strategy because nerve fibers, which display immunoreactivity for VIP, are abundant around intracerebral small arteries and arterioles. This 28 amino acid peptide acts as a potent vasodilator when applied topically to intracranial vessels, and it plays an important role

Table 2 Pharmacokinetics and Brain Uptake of Biotinylated Drugs With and Without Avidin-Based Delivery Vectors

Vectors	Brain concentration (%ID/g)	PS (µl min^{-1} g^{-1})	AUC, 0–60 min [%ID min ml^{-1}]
VIPa	0.013 ± 0.002	<0.5	30 ± 1
+OX26/Av	0.119 ± 0.008	0.85 ± 0.06	141 ± 13
DALDA	0.019 ± 0.002	0.84 ± 0.13	23 ± 1
+OX26/SA	0.12 ± 0.01	0.47 ± 0.07	254 ± 25
K7DA	0.040 ± 0.002	0.85 ± 0.04	47.5 ± 3.4
+OX26/SA	0.14 ± 0.01	0.39 ± 0.05	363 ± 43
BDNF	0.027 ± 0.005	0.67 ± 0.14	13.4 ± 0.8
+OX26/NLA	0.068 ± 0.006	1.70 ± 0.26	39.2 ± 5.7
BDNF-PEG^{2000}	0.021 ± 0.004	<0.05	635 ± 79
+OX26/SA	0.144 ± 0.004	2.0 ± 0.2	74 ± 7
ODN	0.022 ± 0.003	0.57 ± 0.06	39.6 ± 1.6
+OX26/NLA	0.045 ± 0.002	0.58 ± 0.03	77.4 ± 6.4
PS-ODN	0.018 ± 0.002	0.049 ± 0.002	372 ± 11 (2 h)
+OX26/SA	0.041 ± 0.001	0.173 ± 0.006	238 ± 8 (2 h)
PNA	0.0031 ± 0.0002	0.10 ± 0.01	31.2 ± 0.4
+OX26/SA	0.088 ± 0.013	0.61 ± 0.06	143 ± 11
Aβ$_{1-40}$	0.0089 ± 0.0008	—	—
+OX26/SA	0.15 ± 0.01	0.50 ± 0.09	309 ± 43 (2 h)
Aβ$_{1-40}$ (monkey)	<0.15	<0.25	6.00 (3 h)
+HIRMAb/SA	0.62	1.74	3.54 (3 h)

Abbreviations: Aβ$_{1-40}$, amyloid β-peptide; BDNF, brain-derived nemotrophic factor; HIRMAb, human insulin receptor monoclonal antibody; K7DA, [Lys^7]dermorphin analogue; PEG^{2000}, poly(ethylene glycol), MW 2000, PNA, peptide nucleic acid; PS-ODN, phosphorothiorate oligo-deoxynucleotide.
Source: Ref. 135, compiled from data published in Refs. 116, 118, 120, 123, 128, 129, 132, 136.

in the modulation of cerebral blood flow (CBF). Because the receptors are localized on the smooth muscle cells beyond the blood–brain barrier, however, effects on CBF after systemic administration of VIP are not seen. A metabolically stabilized analogue of VIP was constructed that could be biotinylated at a single site under retention of receptor binding and biological activity. Coupling of the biotinylated VIP analogue to the OX26-avidin vector resulted in brain delivery and the desired pharmacological effect: Significant increases in CBF of 65% could be demonstrated after systemic administration of the chimeric peptide. The effect was seen both in anesthetized rats under controlled

Table 3 Pharmacologic Effects Obtained with Chimeric Peptides in Animal Models

Chimeric peptide	Dose	Mode of administration	Animal model	Effect	Ref.
Biotinylated VIP analogue linked to OX26-Av	12 µg/kg	Intracarotid infusion	Rat; artificial ventilation under nitrous oxide anesthesia	Increase in CBF	116
Biotinylated VIP analogue linked to OX26-SA	20 µg/kg or 100 µg/kg	Single iv injection	Rat; conscious	Dose-dependent increase in CBF	119
NGF chemically conjugated to OX26	6.2 µg/injection	Iv injection 4 times every 2 weeks	Rat; intraocular forebrain transplant	Survival of cholinergic neurons	111
NGF chemically conjugated to OX26	50 µg/injection	Iv injection, twice weekly for 6 weeks	Aged rat (24 months)	Improvement of spatial memory in impaired rats	124
NGF chemically conjugated to OX26	20 µg/injection	Iv injection daily for 3 days, then every 2 days, six times	Rat; quinolinic acid lesion	Rescue of striatal cholinergic neurons	125
NGF chemically conjugated to antiprimate TfR mAb AK30		Iv injection	Nonhuman primate	Up-regulation of p75 NGF-receptor in striatum	126
GDNF chemically conjugated to OX26	5 µg/injection	Iv injection 3 times every 2 weeks	Rat; intraocular spinal cord transplant	Survival of motor neurons	127
Biotinylated PEG-BDNF linked to OX26-SA	250 µg/kg	Iv injection daily for 7 days	Rat; transient forebrain ischemia	Rescue of CAl hippocampal neurons	123

Abbreviations: BDNF, brain-derived neurotrophic factor; GDNF, glial cell line derived neurotrophic factor; NGF, nerve growth factor; VIP, vasoactive intestinal polypeptide.

ventilation after intracarotid infusion and in conscious animals after intravenous bolus injection. When an equal dose of the peptide alone without vector was injected (12 µg/kg for the intracarotid infusion or 20 µg/kg in the intravenous study) there was no measurable effect on CBF compared to control animals, but the well-established peripheral effects of VIP on glandular blood flow in the thyroid or the salivary gland were readily detectable. On the other hand, the effect on salivary gland blood flow was attenuated in animals treated with the chimeric peptide delivery system. Taking influence on salivary gland blood flow in that respect as an untoward effect, drug delivery of the VIP analogue to the brain not only resulted in a desired pharmacological response at the target site, it also increased the therapeutic index by diminishing the effect at a nontarget site (119).

The other demonstrations of pharmacological effects of chimeric peptides have been achieved in models of neurodegenerative diseases with different neurotrophic factors. The initial report by Friden et al. used nerve growth factor (NGF) chemically coupled to the vector OX26 via a disulfide linker: an ocular graft model of fetal midbrain placed into the anterior eye chamber of adult rats served to demonstrate the survival-promoting effect of NGF on cholinergic neurones within the grafted tissue (111). Repeated biweekly intravenous injections of the chimeric peptide NGF-OX26 were effective in rescuing the grafts. Further proof of pharmacological effect of the same conjugate was obtained in aged rats with spatial learning deficits. They responded to subchronic treatment with improved performance in the learning task (Morris water maze), and immunohistochemistry showed increased cell size of cholinergic neurons in the medial septal area (124).

The NGF-OX26 chimeric peptide was also effective in a quinolinic acid lesioning model of Huntington's disease (125). Treatment for 2 weeks significantly reduced the loss of intrastriatal cholinergic neurons induced by stereotaxic injection of quinolinic acid. The analogous approach to NGF delivery with an anti–primate transferrin receptor monoclonal antibody, 128.1, has been used in nonhuman primates and has provided immunohistochemical evidence of induction of the p75 NGF receptor in cholinergic striatal neurons (126).

Based on reports showing beneficial effects of local administration of glial cell line derived neurotrophic factor (GDNF) on dopaminergic midbrain neurons in animal models of Parkinson's disease, a conjugate of GDNF with OX26 was studied in a neural graft model (127). The vector-mediated delivery of a dose equivalent to 5µg of GDNF, given as an intravenous bolus three times every 2 weeks, significantly promoted the survival of ocular implants of fetal spinal cord neurons.

Like vector itself, the drug moiety can profoundly influence the pharmacokinetic fate of chimeric peptides. The importance of such effects becomes obvious for drug candidates like the neurotrophins, which are basic peptides: a comparison of the corresponding data in Tables 1 and 2 shows that BDNF decreases the AUC of a chimeric peptide with OX26-NLA as the vector to a level as low as found for the cationic OX26-avidin. Modification of BDNF by introduction of PEG reduced systemic clearance of both the free protein and the conjugate with OX26-SA. The "pegylated" BDNF could be delivered through the BBB by vector-mediated transport as efficiently as the OX26 antibody itself (120). The pharmacokinetically optimized chimeric peptide was used to show the pharmacological potential of BDNF for the treatment of ischemic brain damage (123). In that study, transient forebrain ischemia in rats was induced by bilateral clamping of the carotid arteries. The animals were treated for one week after the insult with chimeric peptide (biotinylated PEG-BDNF coupled to OX26-SA) at a daily dose of 250 µg/kg. Control groups received buffer injections, the vector alone, or the peptide without vector. Brains were histologically processed and the neurons in hippocampal CA1, CA3, and CA4 regions were counted as a morphologic readout. The delivery of BDNF by the vector could fully prevent the neuronal loss in the ischemia-sensitive CA1 region, which amounted to a 68% reduction in cell numbers in the untreated control group. Other controls received an equivalent dose of BDNF alone without vector, which was not neuroprotective.

4. Additional Targets of Vector-Mediated Drug Delivery

Chimeric peptides may be useful for carrying radiopharmaceuticals across the BBB, either for diagnostic or therapeutic purposes. As a diagnostic tool, radiolabeled amyloid peptide $A\beta_{1-40}$ can be delivered to the brain (128, 129). This cleavage product of the amyloid precursor protein (APP) deposits specifically on preexisting amyloidotic plaques and vascular amyloid, which are the hallmarks of Alzheimer's disease (130). Pharmacokinetic studies have been performed in rats with OX26 as a vector (128) and in rhesus monkeys with the insulin receptor antibody 83-14 as a vector (129). In both models radiolabeled $[^{125}I]A\beta$ accumulated in the brain only after vector-mediated delivery. In the monkeys, an analysis of brain sections by phosphoimaging quantitation of radioactivity resulted in images comparable to 2-deoxyglucose scans (129).

Oligodeoxynucleotides are another class of highly hydrophilic macromolecular drug candidates that require transcellular delivery. As a drug constituent of chimeric peptides they also have a potential to affect the pharmacokinetics: coupling of a highly anionic phosphodiester ODN to OX26-NLA increases hepatic clearance and limits brain uptake by lowering the AUC (Table 2, Ref. 131). On the other hand, phosphorothioate-modified ODNs (PS-ODNs) show high plasma protein binding, which may contribute to the low BBB transport measured for a PS-ODN/OX26-SA chimeric peptide (Table 2, Ref. 132). In contrast, the neutral peptide backbone of peptide nucleic acids (PNA) makes these compounds good drug candidates for chimeric peptides and allows for a substantial vector-mediated effect on brain delivery (28-fold increase; see Table 2, Ref. 133).

Because antisense ODNs (and gene therapeutics) have intracellular sites of action in the cytoplasm or nucleus, they require yet another transmembrane transport beyond their delivery through the BBB and cellular uptake by the target cell to elicit specific biological effects: eventual release from the endosomal/lysosomal compartment. That may be realized by the incorporation of endosomal release mechanisms into receptor antibody–based DNA delivery vectors, as demonstrated by the generation of fusion proteins containing the translocation domain of bacterial toxins like exotoxin A (134).

IV. CONCLUSIONS

While there are chances for success of local and invasive delivery strategies in the treatment of localized CNS disease, chronic delivery of macromolecular drugs to the CNS requires a set of noninvasive, systemic approaches that must be developed in a multidisciplinary effort. Significant input is required from cell biology and molecular biology to gain detailed knowledge of cellular transport processes at the BBB. The feasibility of chimeric peptides as a physiological vector-mediated approach for the delivery of peptides, proteins, and oligonucleotides could be demonstrated in animal models. Pharmacological effects in the CNS have already been observed for peptide and protein drugs. As steps toward clinical applicability, the humanization of antibodies used in vectors is expected to reduce immunogenicity, and the generation of fusion proteins will facilitate coupling of vector and drugs. Optimization of neuropharmaceuticals and delivery strategies should proceed in parallel, not sequentially.

REFERENCES

1. LA Wade, R Katzman. Rat brain regional uptake and decarboxylation of L-Dopa following carotid injection. Am J Physiol 228:352–359, 1975.

2. F Hefti. Pharmacology of neurotrophic factors. Annu Rev Pharmacol Toxicol 37:239–267, 1997.
3. RG Blasberg, C Patlak, JD Fenstermacher. Intrathecal chemotherapy: Brain tissue profiles after ventriculocisternal perfusion. J Pharmacol Exp Ther 195: 73–83, 1975.
4. H Davson, K Welch, MB. Segal. Secretion of the cerebrospinal fluid. In: The Physiology and Pathophysiology of the Cerebrospinal fluid. London: Churchill Livingstone, 1987, p 201.
5. R Nau, F Sörgel, HW Prange. Pharmacokinetic optimisation of the treatment of bacterial central nervous system infections. Clin Pharmocokinet 35:223–246, 1998.
6. LM DeAngelis. Current diagnosis and treatment of leptomeningeal metastasis. J Neurooncol 38:245–52, 1998.
7. RD Penn SM Savoy, D Corcos, M Latash, G Gottlieb, B Parke, JS Kroin. Intrathecal baclofen for severe spinal spasticity. N Engl J Med 320:1517–1521.
8. YR Lazorthes, BA Sallerin, JC Verdie. Intracerebroventricular administration of morphine for control of irreducible cancer pain. Neurosurgery 37:422–428, 1995.
9. H Thoenen, C Bandtlow, R Heumann. The physiological function of nerve growth factor in the central nervous system: Comparison with the periphery. Rev Physiol Biochem Pharmacol 109:145–178, 1987.
10. IA Ferguson, JB Schweitzer, PF Bartlett, EM Johnson. Receptor-mediated retrograde transport in CNS neurons after intraventricular administration of NGF and growth factors. J Comp Neurol 313:680–692, 1991.
11. CJ Emmett, GR Stewart, RM Johnson, SP Aswani, RL Chan, LB Jakeman. Distribution of radioiodinated recombinant human nerve growth factor in primate brain following intracerebroventricular infusion. Exp Neurol 140:151–160, 1996.
12. F Hefti. Nerve growth factor promotes survival of septal cholinergic neurons after fimbrial transections. J Neurosci 6:2155–2162, 1986.
13. VE Koliatsos, RE Clatterbuck, HJW Nauta, B Knüsel, LE Burton, FF Hefti, WC Mobley, DL Price. Human nerve growth factor prevents degeneration of basal forebrain cholinergic neurons in primates. Ann Neurol 30:831–840, 1991.
14. M Eriksdotter Jönhagen, A Nordberg, K Amberla, L Bäckman, T Ebendal, B Meyerson, L Olson, A Seiger, M Shigeta, E Theodorsson, M Viitanen, B Winblad, LO Wahlund. Intracerebroventricular infusion of nerve growth factor in three patients with Alzheimer's disease. Dement Geriatr Cogn Disord 9:246–257, 1998.
15. F Hefti (ed) Neurotrophic Factors. Handbook of Experimental Pharmacology. Vol 134. Berlin: Springer-Verlag, 1998, pp 1–325.
16. LG Isaacson, BN Saffran, KA Crutcher. Intracerebral NGF infusion induces hyperinnervation of cerebral blood vessels. Neurobiol Aging 11:51–55, 1990.
17. EJ Mufson, JS Kroin, YT Liu, T Sobreviela, RD Penn, JA Miller, JH Kordower. Intrastriatal and intraventricular infusion of brain-derived neurotrophic factor

in the cynomolgus monkey: Distribution, retrograde transport and co-localization with substantia nigra dopamine-containing neurons. Neuroscience 71:179–191, 1996.

18. CE Krewson, ML Klarman, WM Saltzman. Distribution of nerve growth factor following direct delivery to brain interstitium. Brain Res 680:196–206, 1995.

19. CE Krewson, WM Saltzman. Transport and elimination of recombinant NGF during long-term delivery to the brain. Brain Res 727:169–181, 1996.

20. JH Kordower, SR Winn, YT Liu, EJ Mufson, JR Sladek, JP Hammang, EE Baetge, EF Emerich. The aged monkey basal forebrain: Rescue and sprouting of axotomized basal forebrain neurons after grafts of encapsulated cells secreting human nerve growth factor. Proc Natl Acad Sci USA 91:10898–10902, 1994.

21. PF Morrison, DW Laske, H Bobo, EH Oldfield, RL Dedrick. High-flow microinfusion: Tissue penetration and pharmacodynamics. J Physiol 266:R292–R305, 1994.

22. RH Bobo, DW Laske, A Akbasak, PF Morrison, RL Dedrick, EH Oldfield. Convection-enhanced delivery of macromolecules in the brain. Proc Natl Acad Sci USA 91:2076–2080, 1994.

23. DW Laske, RJ Youle, EH Oldfield. Tumor regression with regional distribution of the targeted toxin TF-CRM 107 in patients with malignant brain tumors. Nat Med 3:1362–1368, 1997.

24. J Greenwood, Mechanisms of blood–brain barrier breakdown. Neuroradiology 33:581–586, 1990.

25. L Sztriha, AL Betz. Oleic acid reversibly opens the blood–brain barrier. Brain Res 550:257–262, 1991.

26. MK Spigelman, RA Zappulla, JD Goldberg, SJ Goldsmith, D Marotta, LI Malis, JF Holland. Effect of intracarotid etoposide on opening the blood–brain barrier. Cancer Drug Delivery 1:3207–3211, 1984.

27. EM Cornford, D Young, JW Paxton, GJ Finlay, WR Wilson, WM Pardridge. Melphalan penetration of the blood–brain barrier via the neutral amino acid transporter in tumor-bearing brain. Cancer Res 52:138–143, 1992.

28. S Nag. Role of the endothelial cytoskeleton in blood–brain barrier permeability to protein. Acta Neuropathol 90:454–460, 1995.

29. Z Nagy, M Szabo, I Huttner. Blood–brain barrier impairment by low pH buffer perfusion via the internal carotid artery in rat. Acta Neuropathol 68:160–163, 1985.

30. T Broman, AM Lindberg-Broman. An experimental study of disorders in the permeability of the cerebral vessels (''the blood–brain barrier'') produced by chemical and physicochemical agents. Acta Physiol Scand 10:102–125, 1945.

31. SI Rapoport, M Hori, I Klatzo. Testing of a hypothesis for osmotic opening of the blood–brain barrier. Am J Physiol 223:323–331, 1972.

32. EA Neuwelt, DL Goldman, SA Dahlborg, J Crossen, F Ramsey, S Roman-Goldstein, R Braziel, B Dana. Primary CNS lymphoma treated with osmotic blood–brain barrier disruption: Prolonged survival and preservation of cognitive function. J Clin Oncol 9:1580–1590, 1991.

33. TS Salahuddin, H Kalimo, BB Johansson, Y Olsson. Observations on exudation of fibronectin, fibrinogen, and albumin in the brain after carotid infusion of hyperosmolar solutions. Acta Neuropathol 76:1–10, 1988.

34. TS Salahuddin, BB Johansson, H Kalimo, Y Olsson. Structural changes in the rat brain after carotid infusions of hyperosmolar solutions. An electron microscopic study. Acta Neuropathol 77:5–13, 1988.

35. AW Vorbrodt, AS Lossinsky, DH Dobrogowska, HM Wisniewski. Cellular mechanism of the blood–brain barrier (BBB) opening to albumin–gold complex. Histol Histopathol 8:51–61, 1993.

36. JD Richmon, K Fukuda, FR Sharp, LJ Noble. Induction of HSP-70 after hyperosmotic opening of the blood–brain barrier in the rat. Neurosci Lett 202:1–4, 1995.

37. PJ Robinson, SI Rapoport. Size selectivity of blood–brain barrier permeability at various times after osmotic opening. Am J Physiol 253:R459–R466, 1987.

38. EM Hiesiger, RM Voorhies, GA Basler, LE Lipschutz, JB Posner, WR Shapiro. Opening the blood–brain and blood–tumor barriers in experimental rat brain tumors: The effect of intracarotid hyperosmolar mannitol on capillary permeability and blood flow. Ann Neurol 19:50–59, 1986.

39. B Zunkeler, RE Carson, J Olson, RG Blasberg, H DeVroom, RJ Lutz, SC Saris, DC Wright, W Kammerer, NJ Patronas, RL Dedrick, P Herscovitch, EH Oldfield. Quantification and pharmacokinetics of blood–brain barrier disruption in humans. J Neurosurg 85:1056–1065, 1996.

40. EA Neuwelt, J Minna, E Frenkel, PA Barnett, CI McCormick. Osmotic blood–brain barrier opening to IgM monoclonal antibody in the rat. Am J Physiol 250: R875–R883, 1986.

41. EA Neuwelt, HD Specht, PA Barnett, SA Dahlborg, A Miley, SM Larson, P Brown, KF Eckerman, KE Hellström, I Hellström. Increased delivery of tumor-specific monoclonal antibodies to brain after osmotic blood–brain barrier modification in patients with melanoma metastatic to the central nervous system. Neurosurgery 20:885–895, 1987.

42. EA Neuwelt, PA Barnett, KE Hellstrom, I Hellstrom, CI McCormick, FL Ramsey. Effect of blood–brain barrier disruption on intact and fragmented monoclonal antibody localization in intracerebral lung carcinoma xenografts. J Nucl Med 35:1831–1841, 1994.

43. LL Muldoon, G Nilaver, RA Kroll, MA Pagel, XO Breakefield, EA Chiocca, BL Davidson, R Weissleder, EA Neuwelt. Comparison of intracerebral inoculation and osmotic blood–brain barrier disruption for delivery of adenovirus, herpesvirus, and iron oxide particles to normal rat brain. Am J Pathol 147:1840–1851, 1995.

44. SE Doran, X Dan Ren, AL Betz, MA Pagel, EA Neuwelt, BJ Roessler, BL Davidson. Gene expression from recombinant viral vectors in the central nervous system after blood–brain barrier disruption. Neurosurgery 36:965–970, 1995.

45. G Nilaver, LL Muldoon, RA Kroll, MA Pagel, XO Breakefield, BL Davidson, EA Neuwelt. Delivery of herpesvirus and adenovirus to nude rat intracerebral tumors after osmotic blood–brain barrier disruption. Proc Natl Acad Sci USA 92:9829–9833, 1995.

46. KL Black. Biochemical opening of the blood–brain barrier. Adv Drug Delivery Rev 15:37–52, 1995.

47. KL Black, CC Chio. Increased opening of blood–tumor barrier by leukotriene C4 is dependent on size of molecules. Neurol Res 14:402–404, 1992.

48. T Inamura, KL Black. Bradykinin selectivity opens blood–tumor barrier in experimental brain tumors. J Cereb Blood Flow Metab 14:862–870, 1994.

49. D Aharony, P Dobson. Discriminative effect of gamma-glutamyltranspeptidase inhibitors on metabolism of leukotrien C4 in peritoneal cells. Life Sci 35:2135–2142, 1984.

50. E Sanovich, RT Bartus, PM Friden, RL Dean, HQ Lee, MW Brightman. Pathway across blood–brain barrier opened by the bradykinin agonist, RMP-7. Brain Res 705:125–135, 1995.

51. TF Cloughesy, KL Black, YP Gobin, K Faharani, G Nelson, P Villablanca, F Kabbinavar, F Vineula, CH Wortel. Intra-arterial Cereport (RMP-7) and carboplatin: A dose escalation study for recurrent malignant gliomas. Neurosurgery 44:270–279, 1999.

52. J Ford, C Osborn, T Barton, NM Bleehen. A phase I study of intravenous RMP-7 with carboplatin in patients with progression of malignant glioma. Eur J Cancer 34:1807–1811, 1998.

53. S Nakano, K Matsukado, KL Black. Enhanced cytokine delivery and intercellular adhesion molecule 1 (ICAM-1) expression in glioma by intracarotid infusion of bradykinin analog, RMP-7. Neurol Res 19:501–508, 1997.

54. DM Goldenberg. Future role of radiolabeled monoclonal antibodies in oncological diagnosis and therapy. Semin Nucl Med 19:332–339, 1989.

55. JD Fenstermacher, RG Blasberg, CS Patlak. Methods for quantifying the transport of drugs across brain barrier systems. Pharmacol Ther 14:217–248, 1981.

56. WM Pardridge. Brain capillary endothelial transport of insulin. In: N Simionescu, M Simionescu, eds. Endothelial Cell Dysfunctions. New York, Plenum Press, 1992, pp 347–362.

57. JB Fishman, JB Rubin, JV Handrahan, JR Connor, RE Fine. Receptor-mediated transcytosis of transferrin across the blood–brain barrier. J Neurosci Res 18: 299–304, 1987.

58. L Fenart, B Dehouck, MP Dehouck, G Torpier, R Cecchelli. Interactions of lipoproteins with the blood–brain barrier. In: WM Pardridge, ed. Introduction to the Blood–Brain Barrier. Cambridge: Cambridge University Press, 1998, pp 221–226.

59. WA Banks, AJ Kastin, W Huang, JB Jaspan, L Maness. Leptin enters the brain by a saturable system independent of insulin. Peptides 17:305–311, 1996.

60. WM Pardridge. Receptor-mediated peptide transport through the blood–brain barrier. Endocr Rev 7:314–330, 1986.

61. WM Pardridge, J Eisenberg, J Yang. Human blood–brain barrier insulin receptor. J Neurochem 44:1771–1778, 1985.

62. JF Haskell, E Meezan, DJ Pillion. Identification and characterization of the insulin receptor of bovine retinal microvessels. Endocrinology 115:698–704, 1984.

63. DJ Pillion, JF Haskell E Meezan. Cerebral cortical microvessels: An insulin sensitive tissue. Biochem Biophys Res Commun 104:686–692, 1982.

64. J Albrecht, B Wroblewska, MJ Mossakowski. The binding of insulin to cerebral capillaries and astrocytes of the rat. Neurochem Res 7:489–494, 1982.

65. DW Miller, BT Keller, RT Borchardt. Identification and distribution of insulin receptors on cultured bovine brain microvessel endothelial cells: Possible function in insulin processing in the blood–brain barrier. J Cell Physiol 161:333–341, 1994.

66. KR Duffy, WM Pardridge. Blood–brain barrier transcytosis of insulin in developing rabbits. Brain Res 420:32–38, 1987.

67. RG Rosenfeld, H Pham, B Keller, RT Borchardt, WM Pardridge. Demonstration and structural comparison of receptors for insulin-like growth factor-I and -II (IGF-I and -II) in brain and blood–brain barrier. Biochem Biophys Res Commun 149:159–166, 1987.

68. KR Duffy, WM Pardridge, RG Rosenfeld. Human blood–brain barrier insulin-like growth factor receptor. Metabolism 37:136–140, 1988.

69. RR Reinhardt, CA Bondy. Insulin-like growth factors cross the blood–brain barrier. Endocrinology 135:1753–1761, 1994.

70. WA Jefferies, MR Brandon, SV Hunt, AF Williams, KC Gatter, DY Mason. Transferrin receptor on endothelium of brain capillaries. Nature 312:162–163, 1984.

71. WM Pardridge, J Eisenberg, J Yang. Human blood–brain barrier transferrin receptor. Metabolism 36:892–895, 1987.

72. S Skarlatos, T Yoshikawa, WM Pardridge. Transport of [^{125}I]transferrin through the rat blood–brain barrier. Brain Res 683:164–171, 1995.

73. A Crowe, EH Morgan. Iron and transferrin uptake by brain and cerebrospinal fluid in the rat. Brain Res 592:8–16, 1992.

74. L Descamps, MP Dehouck, G Torpier, R Cecchelli. Receptor-mediated transcytosis of transferrin through blood–brain barrier endothelial cells. Am J Physiol 270:H1149–H1158, 1996.

75. RL Roberts, RE Fine, A Sandra. Receptor-mediated endocytosis of transferrin at the blood–brain barrier. J Cell Sci 104:521–532, 1993.

76. J Huwyler, WM Pardridge. Examination of blood–brain barrier transferrin receptor by confocal microscopy of unfixed isolated rat brain capillaries. J Neurochem 70:883–886, 1998.

77. U Bickel, Y-S Kang, T Yoshikawa, WM Pardridge. In vivo demonstration of subcellular localization of anti–transferrin receptor monoclonal antibody–colloidal gold conjugate within brain capillary endothelium. J Histochem Cytochem 42:1493–1497, 1994.

78. PL Golden, TJ Maccagnan, WM Pardridge. Human blood–brain barrier leptin receptor. J Clin Invest 99:14–18, 1997.

79. C Bjorbaek, JK Elmquist, P Michl, RS Ahima, A van Bueren, AL McCall, JS Flier. Expression of leptin receptor isoforms in rat brain microvessels. Endocrinology 139:3485–3491, 1998.

80. RJ Boado, PL Golden, N Levin, WM Pardridge. Up-regulation of blood–brain barrier short-form leptin receptor gene products in rats fed a high fat diet. J Neurochem 71:1761–1764, 1998.

81. D Triguero, JB Buciak, WM Pardridge. Capillary depletion method for quantifying blood–brain barrier transcytosis of circulating peptides and plasma proteins. J Neurochem 54:1882–1888, 1990.

82. AW Vorbrodt. Ultracytochemical characterization of anionic sites in the wall of brain capillaries. J Neurocytol 18:359–368, 1989.

83. D Triguero, JB Buciak, J Yang, WM Pardridge. Blood–brain barrier transport of cationized immunoglobulin G: Enhanced delivery compared to native protein. Proc Natl Acad Sci USA 86:4761–4765, 1989.

84. RD Broadwell, BJ Balin, M Salcman. Transcytosis of blood-borne protein through the blood–brain barrier. Proc Natl Acad Sci USA 85:632–636, 1988.

85. WM Pardridge, D Triguero, J Buciak. Transport of histone through the blood–brain barrier. J Pharmacol Exp Ther 251:821–826, 1989.

86. WM Pardridge, JL Buciak, T Yoshikawa. Transport of recombinant CD4 through the rat blood–brain barrier. J Pharmacol Exp Ther 261:1175–1180.

87. Y-S Kang, Y Saito, WM Pardridge. Pharmacokinetics of [^{3}H]biotin bound to different avidin analogues. J Drug Targeting 3:159–165, 1995.

88. WM Pardridge, D Triguero, J Buciak, J Yang. Evaluation of cationized rat albumin as a potential blood–brain barrier drug transport vector. J Pharmacol Exp Ther 255:893–899, 1990.

89. M Shimon-Hophy, KC Wadhwani, K Chandrasekaran, D Larson, QR Smith, SI Rapoport. Regional blood–brain barrier transport of cationized bovine serum albumin in awake rats. Am J Physiol 261:R478–R483, 1991.

90. D Triguero, JL Buciak, WM Pardridge. Cationization of immunoglobulin G results in enhanced organ uptake of the protein after intravenous administration in rats and primate. J Pharmacol Exp Ther 258:186–192, 1991.

91. WM Pardridge, Y-S Kang, A Diagne, JA Zack. Cationized hyperimmune immunoglobulins: Pharmacokinetics, toxicity evaluation, and treatment of HIV-infected hu-PBL-scid mice. J Pharmacol Exp Ther 276:246–252, 1996.

92. U Bickel, VMY Lee, WM Pardridge. Pharmacokinetic differences between ^{111}In- and ^{125}I-labeled cationized monoclonal antibody against β-amyloid in mouse and dog. Drug Delivery 2:128–135, 1995.

93. WM Pardridge, Y-S Kang, J Yang, JL Buciak. Enhanced cellular uptake and in vivo biodistribution of a monoclonal antibody following cationization. J Pharm Sci 84:943–948, 1995.

94. T Terasaki, Y Deguchi, H Sato, K Hirai, A Tsuji. In vivo transport of a dynor-

phin-like analgesic peptide, E-2078, through the blood–brain barrier: An application of brain microdialysis. Pharm Res 8:815–820, 1991.

95. AK Kumagai, J Eisenberg, WM Pardridge. Absorptive-mediated endocytosis of cationized albumin and a β-endorphin-cationized albumin chimeric peptide by isolated brain capillaries. Model system of blood–brain barrier transport. J Biol Chem 262:15214–15219, 1987.

96. JF Poduslo, GL Curran. Polyamine modification increases the permeability of proteins at the blood–nerve and blood–brain barriers. J Neurochem 66:1599–1609, 1996.

97. U Bickel, VMY Lee, JQ Trojanowski, WM Pardridge. Development and in vitro characterization of a cationized monoclonal antibody against βA4 protein: A potential probe for Alzheimer's disease. Bioconjug Chem 5:119–125, 1994.

98. WM Pardridge. Peptide Drug Delivery to the Brain. New York: Raven Press, 1991, pp 198–218.

99. U Bickel, T Yoshikawa, WM Pardridge. Delivery of peptides and proteins through the blood–brain barrier. Adv Drug Delivery Rev 10:205–245, 1993.

100. U Bickel. Antibody delivery through the blood–brain barrier. Adv Drug Delivery Rev 15:53–72, 1995.

101. I Tamai, Y Sai, H Kobayashi, M Kamata, T Wakamiya, A Tsuji. Structure–internalization relationship for adsorptive-mediated endocytosis of basic peptides at the blood–brain barrier. J Pharmacol Exp Ther 280:410–415, 1997.

102. SG Adler, H Wang, HJ Ward, AH Cohen, WA Border. Electrical charge: Its role in the pathogenesis and prevention of experimental membraneous nephropathy in the rabbit. J Clin Invest 71:487–499, 1983.

103. WM Pardridge, J Buciak, J Yang, D Wu. Enhanced endocytosis in cultured human breast carcinoma cells and in vivo biodistribution in rats of a humanized monoclonal antibody after cationization of the protein. J Pharmacol Exp Ther 286:548–554, 1998.

104. WM Pardridge, D Triguero, JL Buciak. β-Endorphin chimeric peptides: Transport through the blood–brain barrier in vivo and cleavage of disulfide linkage by brain. Endocrinology 126:977–984, 1990.

105. M Fukuta, H Okada S Iinuma S Yanai, H Toguchi. Insulin fragments as a carrier for peptide delivery across the blood–brain barrier. Pharm Res 11:1681–1688, 1994.

106. D Wu, Y-S Kang, U Bickel, WM Pardridge. Blood–brain barrier permeability to morphine-6-glucuronide is markedly reduced compared to morphine. Drug Metab Dispos 25:768–771, 1997.

107. WM Pardridge. Vector-mediated peptide drug delivery to the brain. Adv Drug Delivery Rev 15:109–146, 1995.

108. WM Pardridge, Y-S Kang, JL Buciak, J Yang. Human insulin receptor monoclonal antibody undergoes high affinity binding to human brain capillaries in vitro and rapid transcytosis through the blood–brain barrier in vivo in the primate. Pharm Res 12:807–16, 1995.

109. Y-S Kang, U Bickel, WM Pardridge. Pharmacokinetics and saturable blood–

brain barrier transport of biotin bound to a conjugate of avidin and a monoclonal antibody to the transferrin receptor. Drug Metab Dispos 22:99–105, 1994.

110.   PM Friden, L Walus, GF Musso, MA Taylor, B Malfroy, RM Starzyk. Anti-transferrin receptor antibody and antibody–drug conjugates cross the blood–brain barrier. Proc Natl Acad Sci USA 88:4771–4775, 1991.

111.   PM Friden, LR Walus, P Watson SR Doctrow, JW Kozarich, C Bäckman, H Bergman, B Hoffer, F Bloom, AC Granholm. Blood–brain barrier penetration and in vivo activity of an NGF conjugate. Science 259:373–377, 1993.

112.   LR Walus, WM Pardridge, RM Starzyk, PM Friden. Enhanced uptake of rsCD45 across the rodent and primate blood–brain barrier following conjugation to anti-transferrin receptor antibodies. J Pharmacol Exp Ther 277:1067–1075, 1996.

113.   T Yoshikawa, WM Pardridge. Biotin delivery to brain with a covalent conjugate of avidin and a monoclonal antibody to the transferrin receptor. J Pharmacol Exp Ther 263:897–903, 1992.

114.   Y-S Kang, WM Pardridge. Use of neutral-avidin improves pharmacokinetics and brain delivery of biotin bound to an avidin–monoclonal antibody conjugate. J Pharmacol Exp Ther 269:791, 1994.

115.   SU Shin, D Wu, R Ramanthan, WM Pardridge, SL Morrison. Functional and pharmacokinetic properties of antibody/avidin fusion proteins. J Immunol 158: 4797–4804, 1997.

116.   U Bickel, T Yoshikawa, EM Landaw, KF Faull, WM Pardridge. Pharmacologic effects in vivo in brain by vector-mediated peptide drug delivery. Proc Natl Acad Sci USA 90:2618–2622, 1993.

117.   U Bickel, S Yamada, WM Pardridge. Synthesis and bioactivity of monobiotiny-lated DALDA: A mu-specific opioid peptide designed for targeted brain delivery. J Pharmacol Exp Ther 268:791–796, 1994.

118.   U Bickel, Y-S Kang, WM Pardridge. In vivo cleavability of a disulfide-based chimeric opioid peptide in rat brain. Bioconjugate Chem 6:211–218, 1995.

119.   D Wu, WM Pardridge. Central nervous system pharmacologic effect in conscious rats after intravenous injection of a biotinylated vasoactive intestinal peptide analog coupled to a blood–brain barrier drug delivery system. J Pharmacol Exp Ther 279:77–83, 1996.

120.   WM Pardridge, D Wu, T Sakane. Combined use of carboxyl-directed protein pegylation and vector-mediated blood–brain barrier drug delivery system optimizes brain uptake of brain-derived neurotrophic factor following intravenous administration. Pharm Res 15:574–580, 1998.

121.   U Bickel, Y-S Kang, K Voigt. Stability of a disulfide linker in a chimeric peptide during transport through endothelial cells. Naunyn Schmiedeberg's Arch Pharmacol 357 (suppl) R81, 1998.

122.   Y Deguchi, A Kurihara, WM Pardridge. Retention of biologic activity of human epidermal growth factor following conjugation to a blood–brain barrier drug delivery vector via an extended poly(ethylene glycol) linker. Bioconjug Chem 10:32–37, 1999.

123. D Wu, WM Pardridge. Neuroprotection with noninvasive neurotrophin delivery to the brain. Proc Natl Acad Sci USA 96:254–259, 1999.

124. C Bäckman, GM Rose, BJ Hoffer, MA Henry, RT Bartus, P Friden, AC Granholm. Systemic administration of a nerve growth factor conjugate reverses age-related cognitive dysfunction and prevents cholinergic neuron atrophy. J Neurosci 16:5437–5442, 1996.

125. JH Kordower, V Charles, R Bayer, RT Bartus, S Putney, LR Walus, PM Friden. Intravenous administration of a transferrin receptor antibody–nerve growth factor conjugate prevents the degeneration of cholinergic striatal neurons in a model of Huntington disease. Proc Natl Acad Sci USA 91:9077–9080, 1994.

126. JH Kordower, EJ Mufson, AC Granholm, B Hoffer, PM Friden. Delivery of trophic factors to the primate brain. Exp Neurol 124:21–30, 1993.

127. DS Albeck, BJ Hoffer, D Quissell, LA Sanders, G Zerbe, AC Granholm. A noninvasive transport system for GDNF across the blood–brain barrier. Neuroreport 8:2293–2298, 1997.

128. Y Saito, J Buciak, J Yang, WM Pardridge. Vector-mediated delivery of ^{125}I-labeled β-amyloid peptide $A\beta_{1-40}$ through the blood–brain barrier and binding to Alzheimer disease amyloid of the $A\beta_{1-40}$ vector complex. Proc Natl Acad Sci USA 92:10227–10231, 1995.

129. D Wu, J Yang, WM Pardridge. Drug targeting of a peptide radiopharmaceutical through the primate blood–brain barrier in vivo with a monoclonal antibody to the human insulin receptor. J Clin Invest 100:1804–1812, 1997.

130. JE Maggio, ER Stimson, JR Ghilardi, CJ Allen, CE Dahl, DC Whitcomb, SR Vigna, HV Vinters, ME Labenski, PW Mantyh. Reversible in vitro growth of Alzheimer disease β-amyloid plaques by deposition of labeled amyloid peptide. Proc Natl Acad Sci USA 89:5462–5466, 1992.

131. Y-S Kang, RJ Boado, WM Pardridge. Pharmacokinetics and organ clearance of a 3'-biotinylated, internally [^{32}P]-labeled phosphodiester oligodeoxynucleotide coupled to a neutral avidin/monoclonal antibody conjugate. Drug Metab Dispos 23:55–59, 1995.

132. D Wu, RJ Boado, WM Pardridge. Pharmacokinetics and blood–brain barrier transport of [^{3}H]-biotinylated phosphorothioate oligodeoxynucleotide conjugated to a vector-mediated drug delivery system. J Pharmacol Exp Ther 276:206–211, 1996.

133. WM Pardridge, RJ Boado, Y-S Kang. Vector-mediated delivery of a polyamide ("peptide") nucleic acid analogue through the blood–brain barrier in vivo. Proc Natl Acad Sci USA 92: 5592–5596, 1995.

134. J Fominaya, W Wels. Target cell-specific DNA transfer mediated by a chimeric multidomain protein. J Biol Chem 271:10560–10566, 1996.

135. U Bickel, Y-S Kang. Use of chimeric peptides in drug delivery to the brain. In: O Paulson, GM Knudsen, T Moos, eds. Brain Barrier Systems. Alfred Benzon Symposium 45. Copenhagen: Munksgaard, 1999.

136. WM Pardridge, Y-S Kang, JL Buciak, J Yang. Transport of human recombinant brain-derived neurotrophic factor (BDNF) through the rat blood–brain barrier in vivo using vector-mediated peptide drug delivery. Pharm Res 11:738–746, 1994.

11

Using Nanoparticles to Target Drugs to the Central Nervous System

Jörg Kreuter
Institut für Pharmazeutische Technologie, Biozentrum-Niederursel, Johann Wolfgang Goethe-Universität, Frankfurt am Main, Germany

Renad N. Alyautdin
I. M. Sechenov Moscow Medical Academy, Moscow, Russia

I. INTRODUCTION

The blood–brain barrier represents a formidable obstacle for a large number of drugs, including the majority of anticancer agents, peptides, and nucleic acids (1, 2). As a consequence, this barrier prevents effective treatment of many severe and life-threatening diseases, such as brain tumors, Alzheimer's disease, Parkinson's disease, and other neurological disorders. The blood–brain barrier is mainly created by the endothelial cells lining the blood vessels in the brain. These cells are connected by tight circumferential intercellular junctions, an arrangement that effectively abolishes aqueous paracellular pathways across the cerebral endothelium and thus prevents the free diffusion of solutes into the brain. The endothelial cells also contain a variety of metabolizing and detoxifying enzymes, such as cytochrome P450 and monoamine oxidases. In addition, they possess a number of outwardly directed efflux pumps, such as P-glycoprotein (Pgp) and multiple organic acid transporter (MOAT), which actively extrude drugs and neurotoxins from the central nervous system (CNS) (3, 4).

Many attempts have been made to overcome the blood–brain barrier and to transport drugs across it. The most frequent and successful attempts entail chemical modification of the drug (5, 6) or the opening of the blood–brain barrier by osmotic methods (7). However, none of these methods can decisively alter the pharmacological profile of the drug and, in the second case, they represent a massive invasive treatment. An alternative strategy is the employment of liposomes to deliver drugs to the brain (8–11). However, all these methods were not able to promote satisfactorily the passage of inherently nonpenetrating drugs in unmodified form through the intact brain blood vessel endothelium.

An alternative approach is the employment of nanoparticles. Nanoparticles are solid colloid particles ranging in size 1 to 1000 nm (1 μm) (12). They consist of macromolecular materials in which the active principle (drug or biologically active material) is dissolved, entrapped, or encapsulated, or to which the active principle is adsorbed or attached. Nanoparticles can be used therapeutically, for example, as drug carriers or as adjuvants in vaccines (12).

II. EMPLOYMENT OF NANOPARTICLES FOR THE DELIVERY OF DRUGS TO THE BRAIN

A. Empty ^{14}C-Labeled Nanoparticles

A number of years ago Tröster et al. (13, 14) demonstrated that the radioactivity of the brain was significantly higher after intravenous injection of ^{14}C-labeled poly(methyl methacrylate) nanoparticles when the particles, instead of being uncoated, were coated with a variety of surfactants. Among the highest brain concentrations were those obtained after coating with polysorbate 80. However, at that time Tröster et al. did not believe that the nanoparticles were taken up by the brain cells or by the brain blood vessel endothelial cells. They rather assumed an enhanced nanoparticle adsorption to the inner surface of the brain blood vessels (13).

Later, using the same particles, Borchard et al. (15) indeed observed an uptake by cultured bovine brain microvessel endothelial cell monolayers of both uncoated and surfactant-coated poly(methyl methacrylate) nanoparticles. Polysorbate 80 again was by far the most rapid and efficient surfactant. Other surfactants, polyoxyethylene 23-lauryl ether and polysorbate 20, led to a slight uptake enhancement, some (poloxamers 184, 188, and 407) to a delayed uptake enhancement, and some (poloxamer 338 and poloxamine 908) to no uptake enhancement at all compared to the control. In general, with the exception of polysorbate 80, the correlation between the brain concentrations in Tröster's in vivo study (13, 14) and the nanoparticle uptake in Borchard's cell cultures (15) was very poor.

B. Dalargin.

Nanoparticles coated with polysorbate 80 then were used for the delivery of the hexapepide dalargin to the brain (1, 2). Dalargin (Tyr-D-Ala-Gly-Phe-Leu-Arg) is a Leu-enkephalin analogue; it contains D-Ala in second position to prevent enzymatic destruction and thus leads to a high plasma stability. This drug, like other enkephalins, exhibits potent analgesic activity following intracisternal injection. However, it shows no action on the central nervous system after systemic administration up to doses of 20 mg/kg (16). Accordingly, dalargin does not penetrate the blood–brain barrier at all, or only in amounts insufficient to induce an antinociceptive effect.

Upon binding to poly(butyl cyanoacrylate) nanoparticles and overcoating with polysorbate 80, however, investigators observed a considerable dose-dependent analgesic effect when they measured the nociceptive threshold by means of the tail-flick test (1, 2). In this test a hot light ray from a quartz projection bulb is focused on the mouse's tail, and the time for tail withdrawal is recorded (Fig. 1). In addition to analgesia, a pronounced Straub effect was visible. These effects were totally inhibited by naloxone pretreatment. None of the following controls achieved any effects, even with higher doses: uncoated nanoparticles, a simple mixture of the three components (dalargin, nanoparticles, polysorbate 80) mixed immediately before injection, the components alone, or a solution of the drug in 1% polysorbate 80.

The foregoing results were later confirmed by Schröder and Sabel (17) using the hot plate test. The maximal effect was observed after about 45 minutes (Fig. 1) in the studies of Kreuter et al. (1) and Alyautdin et al. (2), but at earlier times by Schröder and Sabel (17) and also by us in Frankfurt. This discrepancy seems to be due to different types of mice used. The Straub effect as well as the total prevention of the nociceptive (analgesic) response by naloxone pretreatment strongly indicate that the drug indeed penetrated the blood–brain barrier after binding to the nanoparticles and overcoating with polysorbate 80 and interacted with the opioid receptors. The latter effects also indicate that the antinociceptive action was not caused by reaction with peripheral receptors.

C. Loperamide

To determine whether the foregoing observations with dalargin were a singular event or whether other drugs also may be transported across the blood–brain barrier in a similar way, those studies were repeated with loperamide (18). Loperamide was chosen because, like dalargin, it produces central pharmacological effects but is unable to pass through the blood–brain barrier. However, it differs totally from dalargin in terms of its chemical and physicochemical

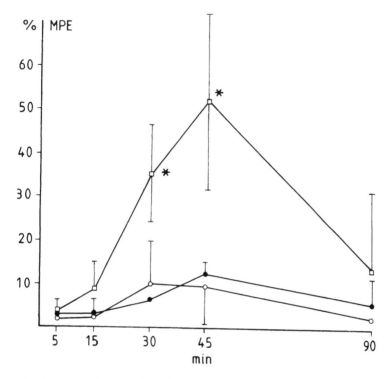

Fig. 1 Antinociceptive effect in percent of the maximally possible effect (% MPE) in mice after intravenous injection of 7.5 mg/kg dalargin adsorbed to poly(butyl cyanoacrylate) nanoparticles and subsequent coating with polysorbate 80 (squares), 10 mg/kg dalargin mixed directly before intravenous injection with poly(butyl cyanoacrylate) nanoparticles and polysorbate 80 (solid circles), or 10 mg/kg dalargin in saline (open circles). (Reprinted from Ref. 1, with permission from Elsevier Science Publishers.)

character. It is not a peptide and unlike dalargin, is poorly water soluble. Neither subcutaneous nor intraperitoneal injection of loperamide in a 10 % propylene glycol solution produced analgesia in the tail-flick test or a Straub effect (19).

Again a strong dose-dependent analgesic effect and a typical Straub effect were observed with loperamide after binding to nanoparticles and overcoating with polysorbate 80, there were no effects without this coating. The mixture of the three components (loperamide, nanoparticles, and polysorbate 80, mixed immediately before injection), as well as a drug solution containing

1% polysorbate 80, produced a 50% smaller and short-lived effect. The production of an aqueous loperamide solution without surfactant was not possible because this drug is only poorly soluble in water. These results confirm the observations with dalargin.

D. Tubocurarine

With a third model drug, tubocurarine, another experimental setup, brain perfusion, was used (20). Tubocurarine, a quaternary ammonium salt, also does not penetrate the normal intact blood–brain barrier. However, the injection of this drug directly into the cerebral ventricles provokes the development of epileptiform seizures as assessed by electroencephalogram (EEG). In the experiments with tubocurarine-loaded nanoparticles the in situ perfused rat brain technique was combined with simultaneous recording of the EEG. Epileptiform spikes appeared in the EEG 15 minutes after the introduction of tubocurarine-loaded polysorbate 80 nanoparticles into the perfusate. Intraventricular injection of tubocurarine caused the appearance of the EEG seizures 5 minutes after administration. Neither a tubocurarine solution nor tubocurarine-loaded nanoparticles without polysorbate 80 or a mixture of polysorbate 80 and tubocurarine were able to influence the EEG. Thus only the loading of tubocurarine onto the polysorbate 80 coated nanoparticles appears to enable the transport of this quaternary ammonium compound through the blood–brain barrier.

E. Doxorubicin

Another study measured whole-brain concentrations of doxorubicin, one of the most important anticancer drugs, as well as the concentrations in the blood, heart, liver, spleen, and lungs. Four preparations were employed: a doxorubicin solution in saline, a doxorubicin solution in saline containing 1 % polysorbate 80, doxorubicin bound to poly(butyl cyanoacrylate) nanoparticles without coating, and, doxorubicin bound to poly(butyl cyanoacrylate) nanoparticles overcoated with polysorbate 80. No difference in blood and all organ concentrations was observed between the solutions with and without polysorbate 80. In the blood, doxorubicin disappeared in a biphasic fashion with all four preparations. Nanoparticles overcoated with polysorbate 80 achieved the highest total doxorubicin blood concentrations, followed by the solutions (Table 1). The lowest concentrations were obtained by doxorubicin bound to uncoated nanoparticles. This was not unexpected, since surfactant coating keeps the particles in blood circulation for an extended time (13, 14), whereas uncoated particles are rapidly taken up by the reticuloendothelial system, which includes

Table 1 Doxorubicin Concentrations in Plasma After Intravenous Injection into Rats, Determined by HLPC

Doxorubicin[a]	Concentrations (µg/ml) in plasma after					
	10 min	1 h	2 h	4 h	6 h	8 h
DX	2.40 ± 0.09	0.41 ± 0.08	0.17 ± 0.08	0.07 ± 0.01	0.04 ± 0.01	0.03 ± 0.01
DX-80	2.70 ± 0.12	0.37 ± 0.04	0.15 ± 0.02	0.05 ± 0.01	0.05 ± 0.01	0.03 ± 0.00
DX-NP	1.68 ± 0.07	0.21 ± 0.03	0.10 ± 0.04	0.08 ± 0.01	0.09 ± 0.01	0.07 ± 0.00
DX-NP-80	3.15 ± 0.42	0.89 ± 0.10	0.21 ± 0.07	0.09 ± 0.01	0.09 ± 0.01	0.06 ± 0.08

[a] DX, doxorubicin [5 mg/kg] in saline; DX-80, doxorubicin [5 mg/kg] in 1% polysorbate 80/saline solution; DX-NP, doxorubicin [5 mg/kg] bound to nanoparticles; and DX-NP-80, doxorubicin [5 mg/kg] bound to nanoparticles and overcoated with 1% polysorbate 80.

Source: Reprinted from J Kreuter, In: AA Hincal and HS Kas, eds. Biomedical Science and Technology. Plenum Press, New York and London, 1998, p 36, with permission from Plenum Publishing Corp., New York.

Table 2 Doxorubicin Concentrations in Heart After Intravenous Injection into Rats, Determined by HPLC

Doxorubicin[a]	Concentrations (μg/g) in heart						
	10 min	1 h	2 h	4 h	6 h	8 h
DX	0.74 ± 0.06	4.27 ± 0.20	6.57 ± 0.55	4.00 ± 0.37	3.30 ± 0.24	1.10 ± 0.05
DX-80	1.10 ± 0.35	8.40 ± 0.62	5.73 ± 0.50	2.10 ± 0.30	0.75 ± 0.08	0.40 ± 0.05
DX-NP	0.28 ± 0.04	0.30 ± 0.05	<0.1	<0.1	<0.1	<0.1
DX-NP-80	0.42 ± 0.07	<0.1	<0.1	<0.1	<0.1	<0.1

[a] Abbreviations as in Table 1.
Source: Reprinted from J Kreuter. In: AA Hincal and HS Kas, eds. Biomedical Science and Technology. Plenum Press, New York and London. 1998, p 36, with permission from Plenum Publishing Corp., New York.

the liver, the spleen, and the lungs. Accordingly, in these organs the concentrations of the doxorubicin were highest with the uncoated nanoparticles. In the heart, an organ in which doxorubicin exhibits severe side effects, the concentration was reduced drastically by both nanoparticle preparations, achieving very low levels that were below the detection limit after 120 minutes (Table 2). The most important results, however, were observed in the brain. Very high concentrations (>6 µg/g organ) were obtained only with the nanoparticles overcoated with polysorbate 80 between 120 and 240 minutes (Fig. 2), whereas all three other preparations were below the detection limit (<0.1 µg/g) all the time. This represents at least a 60-fold difference. The significance of the brain concentrations when nanoparticles coated with polysorbate 80 were used is demonstrated by the observation that the liver levels were only three times higher. The experimental design in this study (removal of entire brain and determination by means of high performance liquid chromatography of the drug after extraction of the brain homogenate) does not discriminate between drug that has been transported across the blood–brain bar-

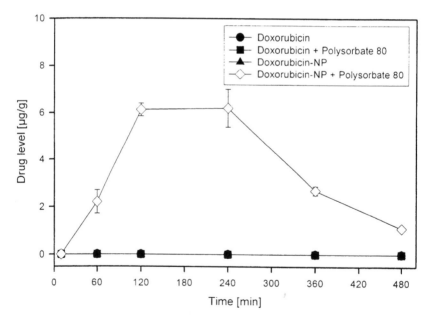

Fig. 2 Uptake of doxorubicin after intravenous injection to rats in the form of one of the four preparations indicated in the inset.

rier and drug that remains in the blood capillaries. However, the blood vessels represent only 1% of the brain volume. Therefore, the observed large brain concentrations as well as the concentration differences to the control preparations taken together with the above-noted results with the other drugs are a strong indication that the nanoparticles indeed enabled a transport of doxorubicin into the brain.

The employment of polysorbate 80 coated nanoparticles loaded with this drug or other anticancer agents thus could lead to new possibilities for the treatment of brain tumors. These tumors are among the most incurable of neoplastic diseases and, therefore, have a bad prognosis.

III. MECHANISM OF NANOPARTICLE-MEDIATED TRANSPORT OF DRUGS ACROSS THE BLOOD–BRAIN BARRIER

The mechanism by which the nanoparticles facilitate drug entry into the brain is not fully elucidated. In principle, six different possibilities exist that could enable enhanced transport of drugs across the blood–brain barrier by means of nanoparticles:

1. Preferential adsorption of nanoparticles to the walls of the brain blood vessel without transport of particles across the endothelium
2. Fluidization of the endothelium by the surface activity of the surfactant polysorbate 80
3. Opening of the tight junctions of the brain endothelium
4. Endocytosis by the brain endothelial cells
5. Transcytosis through the brain endothelial cells
6. Blockage of the P-glycoprotein by polysorbate 80

A. Adhesion of Nanoparticles to Brain Blood Vessel Walls

Tröster et al. (13) suggested that the adhesion of nanoparticles to the inner surface of the brain blood vessels might explain the observed higher brain radioactivity levels with polysorbate 80 coated, ^{14}C-labeled nanoparticles. Adhesion of nanoparticles to another biological surface, the cornea and conjunctiva of rabbit eyes, had been reported previously (21, 22) and indeed led to enhanced delivery into the eye of some ophthalmic drugs (23–27). However, the increase in eye drug transport was very minimal in comparison to the above-reported drug transport enhancement observed in the brain. Additionally, in some cases ophthalmic drug transport was even reduced by binding to nanoparticles (28, 29). In addition, the shear forces that would remove adhering nanoparticles from the biological surface are probably much higher in the brain blood vessel than in rabbit eyes, especially since these animals blink very infrequently. Taking these observations together, it seems very unlikely that simple adhesion of nanoparticles to the brain blood vessel walls would lead to the observed significant drug transport into the brain.

B. Fluidization of the Endothelium by Surfactants

The possibility that the enhanced drug transport is due to the surface activity of polysorbate 80 and a resulting fluidization of the endothelium can be tested by the employment of other surfactants.

Experiments with such surfactants using the tail-flick test and dalargin as the experimental drug showed that polysorbates 20, 40, and 60 also were able to transport dalargin into the brain and produced an antinociceptive effect, although this effect was somewhat lower than with polysorbate 80 (Table 3). However, other surfactants that are listed in Table 3, as well as polysorbates 83 and 85, did not induce any significant antinociceptive effect. This demonstrates that the transport of dalargin and the other drugs across the blood–brain barrier cannot be due to a simple surfactant effect on the endothelial cells or on the tight junctions, since most of the surfactants were inactive. It might be argued that the polysorbates would interact much more strongly with the nanoparticles than other surfactants (30). In fact, the opposite is the case: poloxamine 908 interacts much more strongly with the nanoparticles and influences the body distribution at much lower concentrations than does polysorbate 80 after intravenous injection (31).

C. Opening of the Tight Junctions of the Endothelium

Another possible explanation of the enhanced transport of drugs across the blood–brain barrier is opening of the tight junctions between the endothelial cells lining the brain blood vessels. These junctions may be opened, for instance, by hyperosmotic pressures, enhancing drug transport into the brain (7, 32, 33). This opening can be measured via the inulin space (34–37), which will increase by 10- to 20-fold if significant opening has occurred. The inulin spaces found by Alyautdin et al. (38) after injection of polysorbate 80 coated nanoparticles were increased by much smaller factors (by a factor of 1.1 after 10 min and 2 after 30 minutes) (Fig. 3). In general they were similar to the

Table 3 Mean Percentage of Maximally Possible Effect (MPE) and Standard Deviation (SD) of Nociceptive Threshold After Intravenous Injection of Surfactant Coated and Dalargin-Loaded (10 mg/kg) Poly(butylcyanoacrylate) Nanoparticles into Mice Determined by the Tail-Flick Test

Surfactant (1% solution)	%MPE ± SD after[a]			
	15 min	30 min	45 min	90 min
Polysorbate 20	**79.7** ± 21.3	**51.4** ± 19.0	**52.9** ± 20.9	30.3 ± 30.2
Polysorbate 40	**87.5** ± 16.1	**60.6** ± 41.1	**60.8** ± 38.0	36.5 ± 15.5
Polysorbate 60	25.9 ± 49.7	30.3 ± 35.8	37.9 ± 36.6	61.5 ± 11.4
Polysorbate 80	**100** ± 0	**88.8** ± 22.3	**81.8** ± 14.9	20.2 ± 22.3
Poloxamer 184	0.9 ± 0.28	0.9 ± 5.4	1.0 ± 2.3	0.4 ± 5.1
Poloxamer 188	8.1 ± 5.9	5.5 ± 5.2	3.3 ± 3.4	4.8 ± 3.7
Poloxamer 388	0.2 ± 0.5	1.4 ± 2.4	1.4 ± 3.9	−0.4 ± 5
Poloxamer 407	4.4 ± 3.9	**8.1** ± 2.9	9.5 ± 5.8	6.6 ± 4.4
Poloxamine 908	−1.3 ± 3.6	3.6 ± 5.2	4.2 ± 5.5	3.8 ± 0.51
Brij 35	−1.6 ± 5.0	1.3 ± 8.5	5.5 ± 10	7.2 ± 8.5
Cremophor EZ	10.9 ± 13.1	11.7 ± 15.1	8.6 ± 8.7	2.4 ± 8.8
Cremophor RH 40	**20.2** ± 11.4	**23.6** ± 16.9	31.8 ± 26.7	10.2 ± 12.3
Solution of dalargin	2.3 ± 4.6	10 ± 9.8	9.3 ± 2.8	2 ± 6.1
Uncoated dalargin-loaded nanoparticles	2.3 ± 1.6	4.1 ± 1.0	3.7 ± 11.7	4.9 ± 1.1

[a] Boldface type indicates significant ($p < 0.05$) antinociceptive effects.
Source: Adapted from Ref. 30, with permission from Elsevier Science Publishers.

comparable vascular spaces obtained for the rat in other studies and species (35–37). The small increase would be consistent with the view that polysorbate 80 coated nanoparticles are not significantly disrupting the blood–brain barrier by a general opening of the tight junctions but are increasing the volume of distribution accessible to the intravascular inulin by other means.

In this context it is important to consider the experiments of Alyautdin et al. (2). If the polysorbate 80 or nanoparticles were significantly disrupting the blood–brain barrier by, for example, relaxing tight junctions between the endothelial cells, then the control experiments of Alyautdin et al. (2)—in which dalargin, polysorbate 80, and nanoparticles are injected together, without preadsorption of the dalargin to the nanoparticles—might be expected to

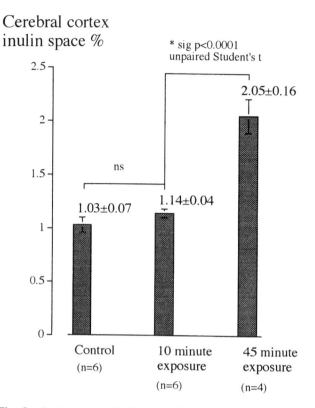

Fig. 3 Cerebral cortex inulin spaces in control rats, and 10 minutes and 45 minutes after the injection of 450 µl of phosphate-buffered saline containing 9 mg of polysorbate 80 coated nanoparticles. Error bars are SEM, and significance is determined by unpaired Student's test.

disrupt the blood–brain barrier sufficiently to admit dalargin (MW 725 Da), to the brain, resulting in analgesia. This does not occur. On the other hand, the rapid onset of the antinociceptive effect observed by Schröder and Sabel (17) and recently also by us with other mouse strains is an indication that a slight opening of the tight junction cannot totally be ruled out at the moment.

D. Endocytosis by the Brain Blood Vessel Endothelial Cells

At present the most likely mechanism for the brain transport of drugs seems to be endocytotic uptake by the endothelial cells lining the brain blood vessels.

These cells belong to the classical reticuloendothelial system (RES) as outlined by Aschoff and are able to endocytose particulate matter under certain circumstances (39). Uptake by these cells of particulate low density lipoproteins (LDL) seems to be a critically important process for the delivery of essential lipids to the brain cells (40). The presence of an LDL receptor has been demonstrated by immunochemistry in rat and monkey brains, and apoliprotein (apo) E and apo A1 containing particles have been detected in human cerebrospinal fluid (40, 41).

In our experiments with poly(butyl cyanocrylate) nanoparticles, an endocytotic uptake of the particles by cultured rat (38), mouse, bovine, and human (unpublished results) brain blood vessel endothelial cells in vitro clearly was observed by laser confocal microscopy after coating with polysorbate 80, but not without this coating. Uptake did not occur at 4°C. In addition, in in vivo experiments FITC-dextran-labeled fluorescent polysorbate 80 coated nanoparticles were visualized by fluorescence microscopy in structures close to the brain blood vessels in mice 45 minutes after intravenous injection, whereas particles without polysorbate 80 appeared to remain in the lumen of the brain blood vessels (1). Structures resembling nanoparticles also were observed by transmission electron microscopy in the brain endothelial cells of these mice. Owing to the tendency of electron microscopy to create artifacts, however, these experiments definitely must be repeated.

Incubation of the poly(butyl cyanoacrylate) nanoparticles in human plasma, followed by washing, removal of the adsorbed plasma components, and analysis of these removed plasma components by two-dimensional polyacrylamide gel electrophoresis demonstrated that the adsorption of apo E on nanoparticles was detectable only if the particles had been coated beforehand with polysorbate 20, 40, 60, or 80 (42). These were exactly the same surfactants that were shown earlier to produce an antinociceptive effect in mice after nanoparticle-loading with dalargin and intravenous injection (30; see Sec. III.B.). Coating with all the other surfactants that did not lead to the antinociceptive effect also did not induce adsorption of apo E (42). These results indicate that precoating of poly(butyl cyanoacrylate) nanoparticles with polysorbate 80, 20, 40, or 60 led to adsorption of apo E and possibly other plasma components. The latter nanoparticles then seem to be able to interact with the LDL receptors on the brain endothelial cells which could lead to their endocytosis by these cells.

The above-mentioned increase in inulin distribution volume (38) could be due to a stimulation of fluid phase endocytosis associated with the internalization of the nanoparticles, a process that also carries inulin into the endothelial cells. The presence of fluorescence-labeled nanoparticles in the cultured brain endothelial cells suggests that the greater space being found by the nanoparticles in vivo is due to a process of cell uptake, with or without subsequent transcytosis, rather than by a partial and selective opening of the endothelial tight junctions.

After endocytosis, delivery of the drug to the other brain cells may occur by desorption of drug from the nanoparticles with or without nanoparticle degradation. The latter is possible, since the polymer employed [i.e. poly(butyl cyanoacrylate)] is very rapidly biodegradable (43, 44). Following its release by desorption or biodegradation, the drug would enter the residual brain by diffusion. Alternatively, transport into these parts of the brain could occur by transcytosis of the nanoparticles with drug across the endothelial cells (40), as discussed in the next section.

E. Transcytosis Across the Brain Endothelial Cells

After the uptake of the nanoparticles by the endothelial cells, the nanoparticles and adsorbed drug may be delivered to the other brain cells by transcytosis of the nanoparticles. Evidence that LDL particles may be transported across the blood–brain barrier by receptor-mediated transcytosis was given by De-houck et al. (40). The observation by this group of nondegradation of LDL in the endothelium indicated that the transcytotic pathway in brain capillary endothelial cells is different from the LDL receptor classical pathway. The observed overcoating of the nanoparticles with apo E discussed above may enable the nanoparticles to use the same pathway as the LDL particles. So far, in work with ^{14}C-labeled poly(methyl methacrylate) nanoparticles, we have not been able to find evidence of transcytosis, either in vitro in the Cec-chelli brain endothelial cell coculture model (40) or in vivo after injection to mice. Perhaps, however, poly(methyl methacrylate) represents a different polymer that does not adsorb plasma proteins of the type necessary for the induction of transcytosis. Unfortunately, dalargin does not bind to the poly (methyl methacrylate) particles. Therefore, we could not use relatively simple methods (e.g., the tail-flick test) to determine whether the methacrylate parti-cles also enabled a drug transport across the blood–brain barrier. Conse-quently, absence of transcytosis with the methacrylate particles is no proof that this may not happen with the poly(butyl cyanoacrylate) nanoparticles. Hence the question of transcytosis with nanoparticles across the brain endothe-lium remains unsolved.

F. Blockage of the P-Glycoprotein in the Brain Endothelial Cells

The sixth possibility for the enhancement of brain transport with the nanopar-ticles could be the inactivation of the P-glycoprotein efflux pump. This glycoprotein is present in the brain endothelial cells (45) and is responsible for multidrug resistance, which represents a major obstacle to cancer chemotherapy (46–48). Surfactants including polysorbate 80 were shown to inhibit this efflux system and to reverse multidrug resistance (46–49). However, as mentioned above, addition of polysorbate to the drug solutions was totally inefficient. On the other hand, this surfactant, of course, may be delivered more efficiently to the brain endothelial cell if it is adsorbed to the nanoparticle surface. This could explain why polysorbate coated on nanoparticles yielded high brain concentrations but not a simple surfactant solution. Nevertheless, we believe that the latter mechanism (i.e., blockage of the efflux system by polysorbte 80) may contribute to the higher doxorubicin uptake mechanism, but induction of the endocytotic uptake may play an equal or much larger role. The reasons for this assumption are the in vitro results described in Sec. III. D, as well as the observation that in contrast to other organs, significant brain concentrations with doxorubicin (see Sec. II.5) were obtained only after 2–4 hours. Such a delayed response seems to be a reflection of the time-consuming process of endocytosis.

IV. CONCLUSIONS

Poly(butyl cyanoacrylate) nanoparticles overcoated with polysorbate 80 or other polysorbates were shown to enable the transport of bound drugs across the blood–brain barrier. Drugs that have successfully been transported across this barrier with the nanoparticles include the hexapeptide dalargin, loperamide, tubocurarine, and doxorubicin. The mechanism of this transport has not yet been fully elucidated. The most probable transport pathway seems to be endocytosis by the blood capillary endothelial cells following adsorption of blood plasma components, most likely apo E, after intravenous injection. These particles could interact with the LDL receptors on the endothelial cells and then be internalized. After internalization by the brain capillary endothelial cells, the drug might be released in these cells by desorption or degradation of the nanoparticles and diffuse into the residual brain. Alternatively, the nanoparticles may be transcytosed across the endothelium as suggested for LDL particles (39). In any case the drug would be able to largely circumvent the efficient efflux pumps (P-glycoprotein, MOAT) that actively transport substances such as drugs out of the endothelial cells and are concentrated in the cell membranes adjacent to the brain capillaries.

In addition to these processes, polysorbates seem to be able to inhibit these efflux pumps. This inhibition could contribute to the brain transport properties of the nanoparticles. Although polysorbates are inefficient as inhibitors of the efflux pumps when used intravenously in the form of a 1% solution, the adsorption of these surfactants to the nanoparticle surface may enhance their concentration in the endothelial cells, thus augmenting the brain transport of simultaneously bound drug.

REFERENCES

1. J Kreuter, RN Alyautdin, DA Kharkevich, AA Ivanov. Passage of peptides through the blood–brain barrier with colloidal polymer particles (nanoparticles). Brain Res 674:171–174, 1995.
2. RN Alyautdin, D Gothier, V Petrov, DA Kharkevich, J Kreuter. Analgesic activity of the hexapeptide dalargin adsorbed on the surface of polysorbate 80-coated poly(butyl cyanoacrylate) nanoparticles. Eur J Pharm Biopharm 41:44–48, 1995.
3. MM Gottesman, I Pastan. Biochemistry of multidrug resistance mediated by multidrug transporter. Annu Rev Biochem 62:682–684, 1993.
4. A Schinkel, E Wagenaar, L van Deemter, CAAM Mol, P Borst. P-glycoprotein in the brain of mice influences the brain penetration of many drugs. J Clin Invest 97:2517–2524, 1996.
5. N Bodor, E Shek, T Higuchi. Delivery of a quaternary pyridium salt across the blood–brain barrier by its dihydropyridine derivative. Science 190:115–156, 1975.
6. A Gilman, LS Goodman, TW Rall, F Murad. Goodman and Gilman's The Pharmacological Basis of Therapeutics. 7th ed. New York: Macmillan, 1985, pp 475–480.
7. EA Neuwelt, PA Barnett. Blood–brain barrier disruption in the treatment of brain tumors. Animal studies. In: EA Neuwelt, ed. Implications of the Blood–Brain Barrier and Its Manipulation. Vol 2, New York: Plenum Press, 1989, pp 195–217.
8. G Toffano, A Leon, D Benvegnu, E Borato, GF Azzone. Effects of brain cortex phospholipids on catecholamine content. Pharmacol Res Commun 8:581–590, 1976.
9. E Bigoa, E Borato, A Bruni, A Leon, G Toffano. Pharmacological effects of phosphatidylserine liposomes: Regulation of glycolysis and energy level in brain. Br J Pharmacol 66:167–174, 1979.
10. ZA Tökes, A Kulcsar St. Peteri, JA Todd. Availability of liposome content to the nervous system. Liposomes and the blood–brain barrier. Brain Res 188:282–286, 1980.
11. K Yagi, M Nai. Glycolipid insertion into liposomes for their targeting to specific organs. In: K Yagi, ed. Medical Application of Liposomes. Tokyo: Japan Scientific Societies Press, Karger, 1986, pp 91–97.
12. J Kreuter. Nanoparticles. In: J Swarbrick, JC Boylan eds. Encyclopedia of Pharmaceutical Technology. Vol 10, New York: Marcel Dekker, 1994, pp 165–190.

13. SD Tröster, U Müller, J Kreuter. Modification of the body distribution of poly(-methyl methacrylate) nanoparticles in rats by coating with surfactants. Int J Pharm 61:85–100, 1990.

14. SD Tröster, J Kreuter. Influence of the surface properties of low contact angle surfactants on the body distribution of ^{14}C-poly(methyl methacrylate) nanoparticles. J Microencapsulation 9:19–28, 1992.

15. G Borchard, KL Audus, F Shi, J Kreuter. Uptake of surfactant-coated poly(-methyl methacrylate) nanoparticles by bovine brain microvessel endothelial cell monolayers. Int J Pharm 110:29–35, 1994.

16. EI Kalenikova, OF Dmitrieva, NN Korobov, SV Zhukova, VA Tischenko. Farmakokinetika dalargina. Vopr Med Khim 34:75–83, 1988.

17. V Schröder, BA Sabel. Nanoparticles, a drug carrier system to pass the blood–brain barrier, permit central analgesic effects of i.v. dalargin injections. Brain Res 674:121–124, 1996.

18. RN Alyautdin, VE Petrov, K Langer, A Berthold, DA Kharkevich, J Kreuter. Delivery of loperamide across the blood–brain barrier with polysorbate 80-coated polybutylcyanoacrylate nanoparticles. Pharm Res 14:325–328, 1997.

19. CJE Niemeggers, FM Lenaerts, PA Janssen. Loperamide (R 18553), a novel type of antidiarrheal agent. Arzneim Forsch (Drug Res) 24:1633–1641, 1974.

20. RN Alyautdin, EB Tezikov, P Ramge, DA Kharkevich, DJ Begley, J Kreuter. Significant entry of tubocurarine into the brain of rats by adsorption to polysorbate 80-coated polybutyl-cyanoacrylate nanoparticles: An in situ brain perfusion study. J Microencapsulation 15:67–74, 1998.

21. RW Wood, VH-K Li, J Kreuter, JR Robinson. Ocular disposition of poly-hexyl-2-cyano[3-^{14}C]acrylate nanoparticles in the albino rabbit. Int J Pharm 23:175–183, 1985.

22. R Diepold, J Kreuter, J Himberg, R Gurny, VHL Lee, JR Robinson, MF Saettone, OE Schnaudigel. Comparison of different models for the testing of pilocarpine eyedrops using conventional eyedrops and a novel depot formulation (nanoparticles). Graefe's Arch Clin Exp Ophthalmol 227:188–193, 1989.

23. R Diepold, J Kreuter, P Guggenbühl, JR Robinson. Distribution of poly-hexyl-2-cyano-[3-^{14}C]acrylate nanoparticles in healthy and chronically inflamed rabbit eyes. Int J Pharm 54:149–153, 1989.

24. A Zimmer, E Mutschler, G Lambrecht, D Mayer, J Kreuter. Pharmacokinetic and pharmacodynamic aspects of an ophthalmic pilocarpine nanoparticle-delivery-system. Pharm Res 11:1435–1442, 1994.

25. AK Zimmer, H Zerbe, J Kreuter. Evaluation of pilocarpine-loaded albumin particles as drug delivery systems for controlled delivery in the eye. I. In vitro and in vivo characterization. J Controlled Release 32:57–70, 1994.

26. AK Zimmer, P Chetoni, MF Saettone, H Zerbe, J Kreuter. Evaluation of pilocarpine-loaded albumin particles as controlled drug delivery systems for the eye. II. Co-administration with bioadhesive and viscous polymers. J Controlled Release 33:31–46, 1995.

27. K Langer, E Mutschler, G Lambrecht, D Mayer, G Troschau, F Stieneker, J Kreuter. Methylmethacrylate sulfopropylmethacrylate copolymer nanoparticles for drug delivery. III. Evaluation as drug delivery system for ophthalmic applications. Int J Pharm 158:219–231, 1997.

28. VHK Li, RW Wood, J Kreuter, T Harmia, JR Robinson. Ocular drug delivery of progesterone using nanoparticles. J Microencapsul 3:213–218, 1986.

29. AK Zimmer, P Maincent, P Thouvenot, J Kreuter. Hydrocortisone delivery to healthy and inflamed eyes using a micellar polysorbate 80 solution or albumin nanoparticles. Int J Pharm 110:211–222, 1994.

30. J Kreuter, VE Petrov, DA Kharkevich, RN Alyautdin. Influence of the type of surfactant on the analgesic effects induced by the peptide dalargin after its delivery across the blood–brain barrier using surfactant-coated nanoparticles. J Controlled Rel 49:81–87, 1997.

31. L Araujo, R Löbenberg, J Kreuter. Influence of the surfactant concentration on the body distribution of nanoparticles. J Drug Target 5:373–385, 1999.

32. EA Neuwelt, M Pagel, PA Barnett, M Glassberg, EP Frenkel. Pharmacology and toxicity of intracarotid Adriamycin administration following osmotic blood–brain barrier modification. Cancer Res 41:4466–4470, 1981.

33. EA Neuwelt, M Glassberg, EP Frenkel. Neurotoxicity of chemotherapeutic agents after blood–brain barrier modification: Neuropathological studies. Ann Neurol 14:316–324, 1983.

34. NJ Abbott, P-O Couraud, F Roux, DJ Begley. Studies on an immortalised brain endothelial cell line: Characterisation, permeability and transport. In: J Greenwood, DJ Begley, MB Segal, eds. New Concepts of a Blood–Brain Barrier. New York: Plenum Press, 1995, pp 239–249.

35. DJ Begley. The interaction of some centrally active drugs with the blood–brain barrier and circumventricular organs. Prog Brain Res 91:163–169, 1992.

36. JE Cremer, MP Seville. Regional brain blood flow, blood volume, and haematocrit values the adult rat. J Cereb Blood Flow Metab 3:254–256, 1983.

37. JD Fenstermacher, P Gross, N Sposito, V Acuff, S Petterrsen, K Gruber. Structural and functional variations in capillary systems within the brain. Ann NY Acad Sci 529:21–30, 1988.

38. RN Alyautdin, A Reichel, R Löbenberg, P Ramge, J Kreuter, DJ Begley. Interaction of poly(butyl cyanoacrylate) nanoparticles with the blood–brain barrier in vivo and in vitro. J Drug Target (submitted).

39. TM Saba. Physiology and pathology of the reticuloendothelial system. Arch Intern Med 126:1031–1052, 1970.

40. B Dehouck, C Fenart, M-P Dehouck, A Pierce, G Torpier, R Cecchelli. A new function for the LDL receptor: Transcytosis of LDL across the blood–brain barrier. J Cell Biol 138:887–889, 1997.

41. RE Pitas, JK Boyles, SH Lee, D Foss, RW Maley. Astrocytes synthesize apolipoprotein E and metabolize apolipoprotein E-containing lipoproteins. Biophys Acta 917:148–161, 1987.

42. M Lück. Plasmaproteinadsorption als möglicher Schlüsselfaktor für eine kontrol-

lierte Arzneistoffapplikation mit partikulären Trägern. Ph. D. thesis, Freie Universität Berlin, 1997, pp 130–154.

43. L Grislain, P Couvreur, V Lenaerts, M Roland, D Deprez-Decampeneere, P Speiser. Pharmacokinetics and distribution of a biodegradable drug-carrier. Int J Pharm 15:335–345, 1983.

44. P Couvreur, L Grislain, V Lenaerts, F Brasseur, P Guiot, A Biernacki. Biodegradable polymeric nanoparticles as drug carrier for antitumor agents. In: P Guiot, P Couvreur, eds. Polymeric Nanoparticles and Microspheres. Boca Raton, FL: CRS Press, 1986, pp 27–93.

45. C Cordon-Cardo, JP O'Brien, D Casals, L Rittmann-Grauner, JL Biedler, MR Melamed, JR Bertino. Multidrug resistance gene (P-glycoprotein) is expressed by endothelial cells at blood–brain barrier sites. Proc Natl Acad Sci. USA 86: 695–698, 1989.

46. DM Woodcock, S Jefferson, ME Linsenmeyer, PJ Crowther, GM Chojnowski, B Williams, I Bertoncello. Reversal of multidrug resistance phenotype with Cremophor EL, a common vehicle for water-insoluble vitamins and drugs. Cancer Res 50:4199–4203, 1990.

47. DM Woodcock, ME Linsenmeyer, GM Chojnowski, AB Kriegler, V Nink, LK Webster, WH Sawyer. Reversal of multidrug resistance by surfactants. Br J Cancer 66:62–68, 1992.

48. T Zordan-Nudo, V Ling, Z Liv, E Georges. Effects of nonionic detergents on P-glycoprotein drug binding and reversal of multidrug resistance. Cancer Res 53:5994–6000, 1993.

49. MM Nerurkar, PS Burto, RT Borchardt. The use of surfactants to enhance the permeability of peptides through Caco-2 cells by inhibition of an apically polarized efflux system. Pharm Res 13:528–534, 1996.

12

Basic CNS Drug Transport and Binding Kinetics In Vivo

Albert Gjedde
Positron Emission Tomography Center, Aarhus General Hospital, Aarhus, Denmark

Antony D. Gee
Addenbrookes Hospital, Cambridge, England

Donald Frederick Smith
Institute for Basic Research in Psychiatry, Psychiatric Hospital, Risskov, Denmark

I. INTRODUCTION

This chapter has two purposes. First, we wish to demonstrate that in vivo imaging methods can be used experimentally to reveal the neurobiology of drug action in living animals and people. Second, because of the complexity of the underlying physiology we wish to caution against simplistic interpretation of the results of such studies.

Numerous factors influence the blood–brain transfer, receptor binding, and metabolism of drugs. Many of these factors can be quantified only in combination by means of labeled tracers. The lumping of factors into combined terms is experimentally advantageous but complicates the interpretation of the results. The conventionally applied lumped terms include the multiple regression coefficients K_1, k_2, k_3, and k_4, as well as p_B, the binding potential, defined as the ratio between k_3 and k_4. With knowledge of the specific activity of the tracer and its concentration in the bloodstream, these estimates permit the determination of blood–brain barrier transfer, metabolic rates, and absolute receptor numbers and affinities. Steady state conditions allow further simplification and concatenation of the lumped terms but additionally obscure the underlying complexity of the possible causes of change. Often several mutually indistinguishable factors may be responsible for the observation of altered experimental variables.

II. BLOOD–BRAIN TRANSFER, BRAIN METABOLISM, AND RECEPTOR BINDING

The pharmacokinetic and -dynamic description of drug action owes much to the physiological and neurochemical description of transfer across the blood–brain barrier, enzyme activity in brain tissue, and the kinetics of receptor association and dissociation.

Drugs active in the central nervous system act by reaching brain tissue and in most cases binding to specific receptors as antagonists or agonists. The transfer of these drugs into the central nervous system is governed by numerous factors. Our knowledge of the influence of several of these factors is incomplete, and this ignorance interferes with the proper understanding of how to design, test, administer, and monitor new drugs. The blood–brain barrier is an important factor but by no means the only one. Distribution volumes in the body, in the circulation, and in brain tissue are equally important and notoriously difficult to assess accurately.

To describe these factors, pharmacologists have proposed and tested numerous mathematical descriptions of tracer and drug pharmacokinetics and pharmacodynamics. The symbols of variables in these descriptions follow a usage which has evolved from the original description of the deoxyglucose method by Sokoloff et al. (1). This usage differs slightly from the conventional symbolism of biophysics.

Because of the profusion of terms, it has been necessary to lump them into measurable entities. The lumping of the terms makes it possible to analyze experimental results numerically, but it also tends to obscure the complexity of the underlying processes. Therefore, the goal of the presentation is not only to explain the origin of the lumped terms but also to warn against simplistic interpretation of experimental results: the current reductionist trend must be balanced by a profound understanding of the underlying kinetics. The most important terms, symbolizing the most influential factors, are introduced and defined in Sec. III.

Constants and time-invariant variables are represented by uppercase symbols. True time variables are represented by lowercase symbols. An exception is the rate constant, symbolized by k. Also, note that the apparent volume, clearance, and extraction fraction symbols ν, κ, and ε are time variables.

III. GLOSSARY OF BASIC VARIABLES

c_a The measured concentration of the drug in arterial whole blood or plasma establishes the driving force for the transfer into the brain tissue across the blood–brain barrier. The concentration usually varies as a function of time after administration of the drug and may be very difficult to measure directly because of binding to plasma proteins and other constituents of whole blood (see Ref 2).

c_v The measured concentration in venous whole blood or plasma is established as the result of several processes, including arterial and tissue concentrations, blood flow, permeability of the blood–brain barrier, and binding in tissue.

c_e The aqueous concentration in tissue of one or more exogenous competitors for binding to receptors determines the degree of competition in the tissue according to the half-saturation affinity constants, K_d, of the exogenous competitors. These competitors can be the unlabeled carrier of a labeled ligand or radioligand when the mass of the carrier is significant. In such cases, the radioligand is no longer a tracer. The symbol χ_e represents the ratio c_e/K_d.

E The "unidirectional" extraction fraction is a constant that signifies the quantity of material transferable to the extravascular tissue, relative to the quantity delivered by the circulation (3),

$$E = 1 - \frac{c_v - c_e}{c_a - c_e} = 1 - \frac{\Delta c_v}{\Delta c_a} \tag{1}$$

where Δ refers to the difference between the whole-blood or plasma concentration and the tissue concentration.

ε When the concentration of the drug in the tissue is negligible ($c_e = 0$), E can be measured as a "first-pass" extraction according to the equation defining the measured variable extraction fraction,

$$\varepsilon = \frac{c_a - c_v}{c_a} \tag{2}$$

This variable extraction fraction is not a constant; when the drug accumulates in the tissue as a function of time, the measured ϵ declines relative to E as shown by equation (1). Thus, only when $c_e(t) = 0$ can we write $E = \epsilon(t)$.

P The permeability of capillary endothelium to the drug is a reflection of passive or facilitated diffusion, driven by a concentration or pressure difference. In the case of gases or fluids, the permeability can also be expressed as a hydraulic conductivity. For fluids, depending on the size of the molecules, more or less of the hydraulic conductivity is established by pores, in or between endothelial cell membranes. When the drug is subject to facilitated equilibrative diffusion by stereoselective membrane transporters, the permeability is only apparent, of course, and must be defined in terms of the Michaelis–Menten constants of the facilitated diffusion. For passive diffusion, P is defined as follows:

$$P = \frac{D}{\bar{x}} \tag{3}$$

where D is the diffusion coefficient and $\bar{x}$ the diffusion distance.

β For gases, the hydraulic conductivity is defined as follows:

$$L = \beta P = \frac{\beta D}{\bar{x}} \tag{4}$$

where β is the aqueous solubility coefficient of a gas in solution, equal to the aqueous concentration achieved for a given partial pressure.

S The surface area through which the diffusion occurs is commonly assumed to be the area of capillary endothelium available for transport of a drug. This area is not necessarily constant. Depending on the experimental situation, many researchers claim that "recruitment" occurs physiologically or pathologically, allowing more capillaries than normal to be perfused. Recruitment then makes a larger area available for diffusion or transport. It is not clear that the surface must be limited to capillaries for all drugs; some of the more diffusible drugs may leave the circulation upstream from capillaries. When the area is the surface of the capillary endothelium, it can be inferred from the capillary density according to the equation (4,5),

$$S = 4\pi rN \tag{5}$$

where the factor 4 arises from the tangled web of the cerebral capillary bed.

F The cerebral blood flow is the flow rate of the blood component in which the solubility coefficient was measured. The flow delivers the drug to the capillary surface area.

α The solvent solubility coefficient of an aqueous solute is the solvent concentration achieved for a given aqueous concentration of a solute, equal to the ratio between the concentration of the solute in the solvent (e.g., whole blood or plasma) relative to the concentration in water (e.g., plasma water). The ratio α was originally calculated by Bohr (3) by the relationship:

$$\alpha = \frac{PS}{F} \ln \left[\frac{\Delta c_a}{\Delta c_v} \right] \tag{6}$$

The magnitude is usually greater than unity, often one or even two orders of magnitude greater. High values of α are a frequent explanation of poor CNS penetration, despite high permeability of the blood–brain barrier. This follows from equations (1)–(6), as solved by Crone (5):

$$E = 1 - e^{-PS/(\alpha F)} \tag{7}$$

However, recent evidence from in vitro and in vivo studies suggests that the unstirred plasma layers in the vicinity of cell membranes cause some drugs bound to plasma proteins to have a higher concentration during transport than at steady state, hence to be in effect "shunted" to the endothelium more rapidly than they would be in simple aqueous solution (2). The novel finding of apparent non-steady-state facilitation of diffusion by plasma proteins introduces situations in which the magnitude of this coefficient is in effect lower than unity, or at least lower than predicted from steady state measurements of α. The reciprocal of α is often referred to as a "free" fraction of the solute, f_1.

V_d The steady state volume of distribution in tissue is the volume of the solvent(s) in which the drug distributes passively, equal to the steady state ratio of the mass of the solute in the tissue to its aqueous concentration. The volume may have several components, including water and other fluids. The ratio between the volume of tissue water in which the solute is dissolved and the volume of distribution is the "free" fraction of the solute in the tissue, $f_2 =$

V_u/V_d, to distinguish it from the free vascular fraction, $f_1 = 1/\alpha$. The ratio V_d/α is often referred to as the partition volume V_e, equal to $V_u f_1/f_2$.

k_{on} The bimolecular constant of association to specific receptors in the brain tissue expresses the likelihood of achieving binding when the drug or ligand is present in the vicinity of receptors. As related to probability, the magnitude is therefore a function of the aqueous concentration of the drug in the vicinity of its receptors.

k_{off} The rate constant of dissociation from the receptors is a function of the average duration of binding of a single molecule.

B_{max} The number of receptors available for binding in the absence of competition is not easy to measure in vivo, since all receptors probably hold some endogenous ligands, including the native neurotransmitters themselves.

c_i The concentrations of one or more endogenous competitors for the binding determines their degree of binding according to association and dissociation constants of their own, defining the half-inhibition affinity constants of the endogenous competitors, K_i. The symbol χ_i represents the ratio c_i/K_i.

IV. MEASURABLE LUMPED VARIABLES

A number of the terms defined in Sec. III cannot be determined in vivo. To overcome this limitation, terms are combined to yield measurable variables. Thus, the fundamental terms yield lumped terms, often in the form of multiple regression coefficients that can be estimated directly in vivo. For example from the ratio V_d/α, defined above as the partition volume V_e, the partition coefficient λ is derived as the ratio $V_e/V_w = f_1/f_2$.

A. Capillary Diffusion Capacity

A major experimentally estimated variable is the capillary diffusion capacity, symbolized by the unidirectional blood–brain clearance K_1. This subsequently defines the fractional tissue washout rate k_2:

$$K_1 = EF = F(1 - e^{-PS/(\alpha F)}) \quad \text{and} \quad k_2 = \frac{\alpha F}{V_d}(1 - e^{-PS/(\alpha F)}) \quad (8)$$

Table 1 Diffusion Limitation of Labeled Water

Condition of ventilation	PaCO$_2$ (partial pressure in arterial blood, kPa)	Water (ml 100 cm^{-3} min^{-1} ± SEM), $n = 11$		
		Unidirectional blood–brain clearance, K_1	Blood flow (butanol), F	Permeability–Surface Area Product, PS
Normocapnia	4.3	39 ± 2	43 ± 3	117 ± 13
Hypercapnia	8.5	78 ± 7	119 ± 6	141 ± 13

The capillary diffusion capacity or clearance is the multiple regression coefficient that accounts for the ease of uptake in the experimental situation. Hence, the unidirectional extraction fraction is also the ratio of K_1 to F. The clearance and the extraction are affected by the three main factors, PS product, blood flow, and plasma solubility coefficient, which often confound the outcome of drug trials. The partition volume V_e is revealed also as the ratio K_1/k_2.

In this laboratory, Smith and coworkers of the Psychiatric Hospital in Risskov, Denmark, used positron emission tomography (PET) of porcine brain to examine the brain uptake and binding of a labeled antagonist of the monoamine transporters, [^{11}C]venlafaxine (Wyeth WY 45.030), an antidepressant with rapid onset of action and efficacy even in treatment-resistant cases (7). In this study, we used positron emission tomography of porcine brain to examine the relationship between cerebral blood flow measured with labeled butanol and the unidirectional clearance of water and venlafaxine to determine their PS products (Tables 1 and 2).

Table 1 shows that the brain uptake of labeled water is almost completely flow-limited at normal blood flow rates; that is, the K_1 of water was almost equal to the rate of blood flow, measured as the K_1 of labeled butanol. However, at higher blood flow rates induced by hypercapnia, the diffusion limitation is revealed as a reduction of the extraction fraction of labeled water from 0.91 to 0.66. The diffusion limitation may be counteracted to some extent by an apparent increase in the PS product of labeled water, which may be caused by increased homogeneity of capillary blood flow rates at the higher average blood flow.

The findings in Table 2 revealed that the uptake of labeled venlafaxine also is almost completely flow-limited at normal blood flow rates. The results raise the possibility that blood flow alterations may be linked to monoamine transporter occupancy as well as to the antidepressant effects of venlafaxine.

Table 2 Diffusion Limitation of Labeled Venlafaxine in Normocapnia (N) and Hypercapnia (H)

	Tracer											
	Butanol		Water					Venlafaxine				
	N	H	N	H				N	H			
Region	K_1	K_1	K_1	K_1	E	PS		K_1	K_1	E	PS	
Thalamus	51	168	45	104	62	162		45	88	52	125	
Brainstem	72	131	70	105	80	212		51	57	74	109	
Frontal cortex	42	115	43	68	73	151		49	84	59	103	
Basal ganglia	42	127	46	85	67	141		43	69	54	100	
Subthalamus	41	150	40	71	53	112		49	79	47	96	
Temporal cortex	43	110	45	78	71	136		39	59	54	85	
Occipital cortex	37	84	37	63	75	116		43	51	61	78	
Cerebellum	34	68	36	52	77	98		36	42	62	65	

B. Binding Potential

Fundamental terms also define the lumped rate constants of association and dissociation of tracer ligands, k_3 and k_4 (8),

$$k_3 = \frac{k_{on}}{V_d}\left(\frac{B_{max}}{1 + \chi_e + \chi_i}\right) \quad \text{and} \quad k_4 = k_{off} \tag{9}$$

where χ_e and χ_i are the ratios C_e/K_d and C_i/K_i. The equation in k_3 is valid only when c_e and c_i can be maintained at constant or negligible levels relative to K_d and K_i.

The binding constants k_3 and k_4 are coefficients of the binding equation,

$$\frac{dm_b^*}{dt} = k_3 m_e^* - k_4 m_b^* \tag{10}$$

where m_b^* is the bound quantity of tracer ligand, and

$$\frac{dm_e^*}{dt} = K_1 c_a^* - (k_2 + k_3)m_e^* + k_4 m_b^* \tag{11}$$

where m_e^* equals $V_d c_e^*$. The differential equations, at equilibrium, define a steady state volume of binding,

$$V_B = \frac{K_1}{k_2}\left(1 + \frac{k_3}{k_4}\right) \tag{12}$$

From the definitions of k_3 and k_4, it follows that their ratio is

$$\frac{k_3}{k_4} = \frac{B_{max}}{K_d V_d(1 + \chi_e + \chi_i)} = \frac{m_b^*(t')}{V_d c_e^*(t')} \tag{13}$$

where m_b^* is the bound quantity of the tracer ligand as before, and c_e^* is the aqueous or "free" concentration of the tracer ligand at the time(s) t' when the binding is in equilibrium $[dm_b^*(t)/dt = 0]$. The $m_b^*/(V_d c_e^*)$ ratio is the "bound-over-free" ratio ("B/F") at binding equilibrium, equal to a binding "potential" (10). The binding potential (p_B) can be determined at equilibrium as follows:

$$p_B = \frac{k_3}{k_4} = \frac{V_B}{V_e} - 1 \tag{14}$$

where V_e is the partition volume (K_1/k_2 ratio), determined in a region of little or no binding (reference region method) or in any region with a nonbinding

Table 3 Binding Potentials of Labeled Venlafaxine

Region	Unidirectional blood–brain clearance, K_1 (ml 100 cm^{-3} min^{-1})	Apparent partition volume, V_B (ml/100 cm^3)	Binding potential [equation (14)], $(V_B/V_e) - 1$ (ratio)
Frontal cortex	50 ± 3	170 ± 74	3.25
Thalamus	53 ± 3	140 ± 87	2.50
Subthalamus	49 ± 4	134 ± 87	2.35
Basal ganglia	53 ± 3	69 ± 23	0.73
Temporal cortex	47 ± 3	65 ± 22	0.72
Occipital cortex	50 ± 2	53 ± 18	0.33
Brainstem	54 ± 4	39 ± 11	0
Cerebellum[a]	38 ± 2	40 ± 14	0

[a] Reference region.

enantiomer (inactive enantiomer method). When C_e and C_i are negligible compared to K_d and K_i (i.e., in the absence of competitors and inhibitors), the standard binding potential, p_{B_0}, equals the $B_{max}/(K_d V_d)$ ratio.

When the exchange between the specifically bound and unbound ligand quantities is sufficiently rapid, the bound and the free ligands occupy a single pool of tracer, of which the clearance time constant is a function of the binding potential,

$$\Theta_B = \frac{1 + p_B}{k_2} = \frac{V_B}{K_1} \tag{15}$$

such that $V_B = \Theta_B K_1$.

The binding potential of venlafaxine was determined in regions of the pig brain from estimates of the steady state partition volumes (Table 3), using the reference region method.

V. EXPERIMENTAL DETERMINATION OF BINDING VARIABLES

A. Single-Compartment Estimates of Reversible Binding

According to the definition of a single compartment, the binding equation for tracers subject to rapidly reversible uptake, solution, or specific or nonspecific binding, is (11).

$$\frac{dm^*}{dt} = V_0 \frac{dc_a^*}{dt} + K_1 c_a^* - \frac{m^*(t)}{\Theta_B} \tag{16}$$

where $m^* = m_e^* + m_b^*$. When integrated, this binding equation yields

$$m^*(T) = V_0 c_a^* + K_1 \int_0^T c_a^* \, dt - \frac{1}{\Theta_B} \int_0^T m^* dt \tag{17}$$

where m^* is the total tracer quantity in the tissue as a function of time. This solution of the binding equation leads to the definition of several time-dependent variables, including the time-dependent apparent volume of accumulated uptake $v^*(T)$:

$$v^*(T) = \frac{\int_0^T m^* dt}{\int_0^T c_a^* dt} \tag{18}$$

the time-dependent apparent vascular clearance $\kappa^*(T)$,

$$\kappa^*(T) = \frac{m^*(T)}{\int_0^T c_a^* dt} \tag{19}$$

and the time-dependent virtual time variable $\theta^*(T)$ (12),

$$\theta^*(T) = \frac{\int_0^T m^* dt}{m^*(T)} \tag{20}$$

which interrelate as $v^* = \theta^* \kappa^*$.

B. The Graphical Reversible Binding Plots

Equation (17) is suitable for multiple regression with three coefficients. To linearize the equation, the number of coefficients must be reduced to two. This can be approximated if the measured volume occupied by the tracer is termed v^* and defined as the ratio m^*/c_a^*. The ratio V_0/v^* is then unity at first but eventually must reach the steady state ratio V_0/V_B, which is much less than unity, depending on the magnitude of V_B. The product $\Theta_B (v^* - V_0)/v^*$ is symbolized by Θ_B' which is close to Θ_B for most times.

With this modification, equation (17), the integral equation solution to the binding equation, can be linearized in three different ways, leading to a plot of either *volume (vs. clearance)* (12):

$$v^* = V_B - \Theta_B' \kappa^* \tag{21}$$

yielding the steady state volume of partition as the ordinate intercept (and the unidirectional clearance K_1 as the abscissa intercept), or the *Logan plot* (13):

$$\theta^* = -\Theta'_B + \frac{V_B}{\kappa^*} \qquad (22)$$

in which the steady state volume of partition is the slope (and $-\Theta_B$ the ordinate intercept), the *clearance* (*vs. volume*) plot (14):

$$\kappa^* = K_1 - \frac{v^*}{\Theta'_B} \qquad (23)$$

yielding the unidirectional clearance as the ordinate intercept.

For the volume plot, the binding potential can be calculated from the ratios of the ordinate intercepts of the uptake in binding and nonbinding regions, or of the uptake of binding and nonbinding enantiomers of the tracer.

All three plots are in principle valid for all tracers distributed in a single compartment in the brain or other organs, including blood flow tracers and water (and oxygen gas).

The volume and Logan plots are illustrated in Fig. 1 for the tracer radioligand [^{11}C]raclopride (15). The clearance plot is illustrated in Fig. 2 for the [^{15}O]water distribution in porcine brain (16).

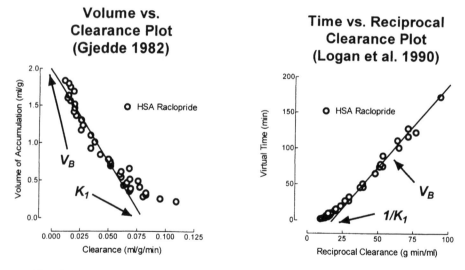

Fig. 1 Volume and Logan plots [see equations (21) and (22), respectively] of the human brain uptake of the tracer radioligand [^{11}C]raclopride. (Redrawn from Ref. 15.)

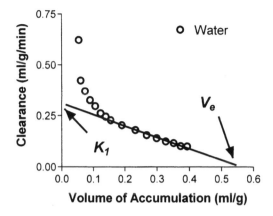

Fig. 2 Clearance plot [see equation (23)] of the porcine brain uptake of [^{15}O]water distribution. (From Ref. 16.)

C. Experimental Determination of Receptor Occupancy

The importance of the binding potential is the ease with which it can be determined in brain imaging experiments. Because the binding potential contributes to the magnitude of the habitual, or apparent, steady state volume of partition, and because the nonspecific binding, defined by the magnitude of V_e, is often assumed to be the same in all regions of the brain, the binding potential can be determined at steady state without recourse to measurements of the radioligand concentration curve in the bloodstream.

The measurements of the binding potential are used to calculate the degree of receptor occupation from the change of binding potential caused by the binding of a "cold" drug, competitor, or inhibitor

$$\sigma_i = \frac{\chi_i}{1 + \chi_e + \chi_i} = 1 - \frac{p_{B_i}}{p_B} \tag{24}$$

where σ_i is the inhibitor occupancy, χ_i the inhibitor concentration relative to the inhibition constant, and p_B and p_{B_i} the binding potentials in the presence of the endogenous ligand alone or in the presence of an exogenous inhibitor as well (17),

$$\chi_i = \sigma_i \frac{p_{B_0}}{p_{B_i}} = p_{B_0} \left[\frac{1}{p_{B_i}} - \frac{1}{p_B} \right] \tag{25}$$

where p_{B_0} is the binding potential of the receptors in the absence of all occupants. Except for this scaling factor, it is possible to monitor changes of extracellular ligand concentrations this way.

In this laboratory, Gee and coworkers at the Aarhus General Hospital used a labeled inhibitor of phosphodiesterase IV, [¹¹C]rolipram, to estimate the number of active sites on this enzyme engaged in second messenger activity (18). The uptake of radiolabeled rolipram was reversible and could be depressed by low specific activity administration, as shown in Figs. 3 and 4.

Figure 3 shows the time–activity curves in pig brain frontal cortex of the active enantiomer (−)-rolipram at high and low specific activity and the inactive enantiomer (+)-rolipram at high specific activity. The inactive enantiomer reveals the partition volume of rolipram in the absence of binding.

Figure 4 shows the volume plots of the three time–activity curves, indicating the steady state volumes of binding of the active enantiomer and the

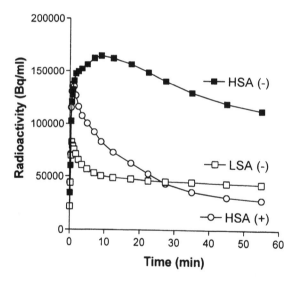

Fig. 3 Time–activity curves in pig brain frontal cortex of the active enantiomer (−)-[¹¹C]rolipram (a phosphodiesterase IV inhibitor) at high and low specific activity and the inactive enantiomer (+)-[¹¹C]rolipram at high specific activity. The inactive enantiomer reveals the partition volume of rolipram in the absence of binding.

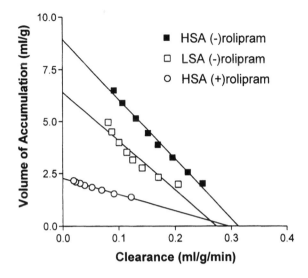

Fig. 4 Volume plots of the porcine brain time–activity curves shown in Fig. 3, indicating the steady state volumes of binding of the active enantiomer and the partition volume of the inactive enantiomer. The regression results are listed in Table 4.

partition volume of the inactive enantiomer, as listed in Table 4. The occupancy of low specific activity rolipram is shown in Table 4 to be 0.4 at a rolipram concentration of about 400 pmol/cm³, calculated according to equation (24), using the inactive enantiomer method.

Table 4 Estimates of Variables of Binding of Radiolabeled Rolipram in Pig Frontal Cortex

Specific activity (Bq/pmol)	Volume of distribution (ml/cm³): equation (21)	Binding potential (ratio): equation (14)	Bound radioligand (pmol/cm³): equation (26)	Free radioligand (pmol/cm³)	Occupancy σ_i (ratio) equation (24)
54,000	8.93	2.95	0.016	0.005	—
38	6.41	1.84	740	400	0.38
Inactive[a]	2.26	—	—	—	—

[a] Inactive enantiomer.

D. Experimental Determination of Receptor Number and Affinity

Multiple measurements of saturation as a function of the "cold" radioligand concentration permit the calculation of the number of receptors B_{max} and the inhibitory constant of the radioligand. According to the Michaelis–Menten equation used earlier, the higher the concentration of ligands, competitors, and inhibitors, the lower the binding potential and saturation. This is also the rationale behind the use of binding potential to make inferences about the concentration of endogenous ligands. The linear equation underlying the Scatchard-like plot can be adapted for use with results of in vivo imaging of radioligand binding,

$$B = \frac{p_B}{1 + p_B}\left(\frac{m^*}{A^*}\right) = B_{max} - K_I V_d\, p_B \qquad (26)$$

where m^* is the radioactivity recorded in a region of interest at steady state (transient equilibrium), A^* the specific radioactivity ratio, and K_I the IC_{50} of the radioligand ($K_i[1 + \chi_e]$).

Figure 5 illustrates the calculation of B_{max} on the basis of equation (26) by means of a Scatchard-like plot. The B_{max} was estimated to be 2000 pmol/

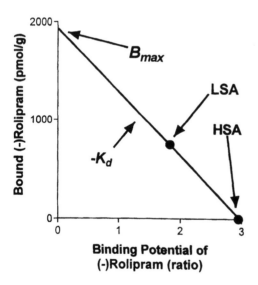

Fig. 5 Scatchard-like plot of bound (–)-rolipram versus binding potential. Binding potential was calculated according to equation (21) from results shown in Fig. 4. Bound (–)-rolipram was calculated according to equation (26) from the binding potential.

cm^3, as indicated by the ordinate intercept and $K_I V_d$ to be 0.7 pmol/cm^3, as indicated by the slope.

REFERENCES

1. L Sokoloff, M Reivich, C Kennedy, MH Des Rosiers, CS Patlak, KD Pettigrew, O Sakurada, M Shinohara. The [^{14}C]deoxyglucose method for the measurement of local cerebral glucose utilization: Theory, procedure, and normal values in the conscious and anesthetized albino rat. J Neurochem 28:897–916, 1977.
2. L Bass, A Gjedde, P Ott. Plasma–cell exchange of protein-bound ligands: Can flux ratios diagnose facilitation? In: OB Paulson et al, eds. Alfred Benzon Symposium 34 Munksgaard, Copenhagen, 1999, pp. 461–472.
3. C Bohr. Über die spezifische Tätigkeit der Lungen bei der respiratorischen Gasaufnahme und ihr Verhalten zu der durch die Alveolarwand stattfindenden Gasdiffusion. Skand Arch Physiol 22:221–280, 1909.
4. ER Weibel. Stereological methods in cell biology: Where are we—where are we going? J Histochem Cytochem 29:1043–1052, 1981.
5. A Gjedde, H Kuwabara, A Hakim. Reduction of functional capillary density in human brain after stroke. J Cereb Blood Flow Metab 10:317–326, 1990.
6. C Crone. The permeability of capillaries in various organs as determined by use of the "indicator diffusion" method. Acta Physiol Scand 58:292–305, 1963.
7. DF Smith, PN Jensen, AD Gee, SB Hansen, EH Danielsen, F Andersen, PA Saiz, A Gjedde. PET Neuroimaging with [^{11}C]Venlafaxine: serotonin uptake inhibition, biodistribution and binding in living pig brain. Eur Neuropsychopharmacol 7:195–200, 1997.
8. A Gjedde, DF Wong. Modeling neuroreceptor binding of radioligands in vivo. In: J Frost, HN Wagner Jr, eds. Quantitative Imaging of Neuroreceptors. New York: Raven Press, 1990, pp 51–79.
9. A Gjedde, DF Wong, HN Wagner Jr. Transient analysis of irreversible and reversible tracer binding in human brain in vivo. In: L Battistin, ed. PET and NMR: New Perspectives in Neuroimaging and Clinical Neurochemistry. New York: Alan R Liss, 1986, pp 223–235.
10. MA Mintun, ME Raichle, MR Kilbourn, GF Wooten, MJ Welch. A quantitative model for the in vivo assessment of drug binding sites with positron emission tomography. Ann Neurol 15:217–227, 1984.
11. A Gjedde. Compartmental analysis. In: HN Wagner Jr, Z Szabo, JW Buchanan, eds. Principles of Nuclear Medicine. 2nd ed., Philadelphia: Saunders, 1995, pp 451–461.
12. A Gjedde. Calculation of glucose phosphorylation from brain uptake of glucose analogs in vivo: A re-examination. Brain Res Rev 4:237–274, 1982.
13. J Logan, JS Fowler, ND Volkow, AP Wolf, SL Dewey, DJ Schlyer, RR MacGregor, R Hitzemann, B Bendriem, SJ Gatley. Graphical analysis of reversible

radioligand binding from time–activity measurements applied to [N-^{11}C-methyl]-(–)-cocaine PET studies in human subjects. J Cereb Blood Flow Metab 10:740–747, 1990.

14. A Gjedde, J Reith, S Dyve, G Léger, M Guttman, M Diksic, AC Evans, H Kuwabara. DOPA decarboxylase activity of the living human brain. Proc Nat Acad Sci USA 88:2721–2725, 1991.

15. DF Wong, T Solling, F Yokoi, A Gjedde. Quantification of extracellular dopamine release in schizophrenia and cocaine use by means of TREMBLE. In: RE Carson, E Daube-Witherspoon, P Herscovitch, eds. Quantitative Functional Brain Imaging with Positron Emission Tomography. San Diego, CA: Academic Press, 1998, pp 463–468.

16. PH Poulsen, DF Smith, L Ostergaard, EH Danielsen, A Gee, SB Hansen, J Astrup, A Gjedde. In vivo estimation of cerebral blood flow, oxygen consumption and glucose metabolism in the pig by [O-15]water injection, [O-15]oxygen inhalation, and dual injections of[F-18]fluorodeoxyglucose. J Neurosci Methods 77:199–209, 1997.

17. M Laruelle, CD D'Souza, RM Baldwin, A Abi-Dargham, SJ Kanes, CL Fingado, JP Seiby, SS Zoghbi, MB Bowers, P Jatlow, DS Charney, RB Innis. Imaging D2 receptor occupancy by endogenous dopamine in humans. Neuropsychopharmacology 17:162–174, 1997.

18. AD Gee, DF Smith, L Ostergaard, D Bender, PH Poulsen, C Simonsen. Mapping second messenger activations in the brain. NeuroImage 7:A38, 1998.

Index